Palgrave Studies in Climate Resilient Societies

Series Editor
Robert C. Brears, Avonhead, Canterbury, New Zealand

The Palgrave Studies in Climate Resilient Societies series provides readers with an understanding of what the terms **resilience and climate resilient** societies mean; the best practices and lessons learnt from various governments, in both non-OECD and OECD countries, implementing climate resilience policies (in other words what is 'desirable' or 'undesirable' when building climate resilient societies); an understanding of what a resilient society potentially looks like; knowledge of when resilience building requires slow transitions or rapid transformations; and knowledge on how governments can create coherent, forward-looking and flexible policy innovations to build climate resilient societies that: support the conservation of ecosystems; promote the sustainable use of natural resources; encourage sustainable practices and management systems; develop resilient and inclusive communities; ensure economic growth; and protect health and livelihoods from climatic extremes.

More information about this series at
http://www.palgrave.com/gp/series/15853

Robert Kijak

Water Asset Management in Times of Climate Change and Digital Transformation

palgrave
macmillan

Robert Kijak
Asset Management & Sustainability
Senglea, Malta

ISSN 2523-8124 ISSN 2523-8132 (electronic)
Palgrave Studies in Climate Resilient Societies
ISBN 978-3-030-79359-3 ISBN 978-3-030-79360-9 (eBook)
https://doi.org/10.1007/978-3-030-79360-9

This Palgrave Macmillan imprint is published by the registered company Springer Nature Switzerland AG
The registered company address is: Gewerbestrasse 11, 6330 Cham, Switzerland

I dedicate this book to my wife, Zofia, who made it possible.

Contents

1 Introduction 1

1.1 Strategic Drivers 2

1.2 Digital Transformation Improving Effectiveness and Efficiency 4

1.3 Water 4.0 Adaptation and Climate Resilience Through Asset Management Systems 5

1.4 Content of This Book 6

References 7

2 Glossary of Major Terms and Acronyms 9

References 22

3 How Can Climate Change Impact upon Water Supply Assets? 27

3.1 A Risk-Based Approach to Climate Change 28

3.2 Flood-Drought Cycle 37

3.3 LOS Impact 44

References 48

4 Design and Operational Considerations for Water Supply Assets 53
4.1 Water Treatment, Transmission and Distribution Assets 58
4.2 Water Storages 64
4.3 Natural Assets 66
References 68

5 Defining Water 4.0 73
5.1 Relationship Between 4.0's 74
5.2 Scope and Timescales of Four Water Revolutions 75
5.3 Technologies and Solutions 78
5.4 Opportunities and Risks 81
References 84

6 Application of Water 4.0 Technologies and Solutions 87
References 121

7 A System for Managing Assets Throughout Their Life 125
7.1 Scope 128
7.2 Implementation 129
7.3 How Water 4.0 Fits into an ISO 55001 Asset Management System? 131
7.4 Water 4.0 Implementation Through ISO 55001 Standard 131
7.5 Digital Twins Transforming Asset Management 133
7.6 Alignment of Financial and Non-Financial Functions 135
7.7 Adaptability or Continuous Improvement 136
7.8 Strategising Information Security 138
7.9 Other Opportunities of the AMS Development 140
References 141

8 Conclusions 145
References 149

References 151

Index 169

List of Figures

Fig. 3.1	An iterative risk management approach in the climate change context	29
Fig. 3.2	Rainfall deficiencies for the period 1 July–31 December 2019 (BOM 2020b) (Reproduced by permission of Bureau of Meteorology, © 2020 Commonwealth of Australia)	37
Fig. 3.3	Rainfall deficiencies for the period 1 November 2005–31 October 2009 (BOM, 2019) (Reproduced by permission of Bureau of Meteorology, © 2020 Commonwealth of Australia)	39
Fig. 3.4	Impacts of turbidity laden water inflow on the upper and lower sections of the Wivenhoe Dam (Watkinson et al., 2012)	43
Fig. 4.1	Risk treatment categories with the current contextual examples (left hand side flowchart: current best practices; right hand side flowchart: potential future practices). The potential future practices have been illustrated in Fig. 4.2	55

Fig. 4.2 An iterative accommodate-retreat/avoid process (Note that it is only the first iteration for the water stressed countries. The left hand side boxes contain the current best practices for water utilities, which are likely to become more common in the near future. The right hand side boxes contain the potential future practices) 57

Fig. 4.3 An illustration of the polishing process for WWTP effluent to produce *new water* in Malta (Water Services Corporation n.d.) 61

Fig. 5.1 Relationship between 4.0's for various industry sectors and the corresponding engineering and business processes 74

Fig. 5.2 Scope and timescales of water revolutions (Vestner & Keilholz, 2016) 77

Fig. 6.1 A screenshot showing example nitrification process values in the automatic blower control mode from the Endress + Hauser's Liquiline Control CDC81 system (Świerczewska, 2020) 89

Fig. 6.2 Upper graph: Example of typical level-based flow control, Lower graph: Example of predictive flow control (Bakker et al., 2013) (Reproduced from JWSRT—AQUA, 2013, volume 62, issue 1, pages 1–13, with permission from the copyright holders, IWA Publishing) 92

Fig. 6.3 BLU-X Wastewater Network Optimization monitoring hydraulic grade lines, CSO outfall and control gates at Evansville 98

Fig. 6.4 Relationship between the main DynaPredict solution components (*Source* http://www.dynamox.net) 103

Fig. 6.5 A screenshot from of the DMA Dashboard (*Source* DynaPredict Web Platform) 104

Fig. 6.6 Spectra collected over time, as indicated on the chronological timeline by the red arrows, and continuous velocity data (a graph below the timeline) 104

Fig. 6.7 Velocity-frequency spectrum showing peaks aligning with indicators at 1X RPM, 2X RPM, 3X RPM etc. (1 RPM = 1/60 Hz) 106

Fig. 6.8	Thermal imaging showing the difference in heat dissipation in the misaligned condition	107
Fig. 6.9	Image showing the component wear after the vibration identification	107
Fig. 6.10	Velocity-frequency spectrum with harmonic indicators at 350 Hz and 700 Hz	108
Fig. 6.11	Velocity-frequency spectrum showing a large spike at 50 Hz	110
Fig. 6.12	Simultaneous increase in vibration levels (upper graph) and temperature (lower graph)	111
Fig. 6.13	Acceleration-frequency spectrum indicating peaks aligning with the BPFO indicators	112
Fig. 6.14	Damage to the bearing	113
Fig. 6.15	Kando staff installing an IoT unit in the sewer manhole (Gelman 2020)	114
Fig. 6.16	An illustration of vDMAs developed by Xylem for a water authority in the United Arab Emirates	120
Fig. 7.1	Approaching the climate resilience goal through Water 4.0 incorporated into AMS	127
Fig. 7.2	Application of the PDCA cycle throughout the asset life cycle	130
Fig. 7.3	Integrated and life cycle organisational management with ISO 9001 (ISO, 2015b), asset management with ISO 55001 (ISO, 2014a) and information management with ISO 19650 (ISO, 2018f) within the physical and digital environments. The original figure 3 from ISO 19650–1: 2018: Organization and digitization of information about buildings and civil engineering works, including building information modelling (BIM)—Information management using building information modelling—Part 1: Concepts and principles, is reproduced with the permission of the International Organization for Standardization, ISO. This standard can be obtained from any ISO member and from the website of the ISO Central Secretariat at the following address: http://www.iso.org. Copyright remains with ISO	134

Fig. 7.4 Increased business value with the gradual Water 4.0 introduction (Weitze et al., 2018) 137
Fig. 7.5 Security-minded management in accordance with ISO 19650–5 (ISO, 2020b) 139

List of Tables

Table 3.1	Summaries of the projected climate change hazards and the adopted resilience/adaptation measures (2014 perspective) for Australia and southern Europe (Hewitson et al., 2014; Kovats et al., 2014; Reisinger et al., 2014, unless otherwise stated/updated in the table)	32
Table 3.2	A summary of weather conditions (rainfall and temperatures only) for Australia and Malta in 2019 and 2020 (BOM 2020a, 2021 respectively). Reproduced by permission of Bureau of Meteorology, © 2020 Commonwealth of Australia. Maltese Islands Weather 2020 and Maltese Islands Weather 2021 respectively	35
Table 3.3	The key meteorological and hydrological conditions leading to the 2011 Queensland Flood	41
Table 3.4	Meteorological and hydrological conditions leading to the 2019 Townsville flood (JBA Group, 2019)	43
Table 3.5	Impacts of contaminants on conventional water treatment processes (Canning et al., 2020)	47

List of Boxes

Box 3.1	Examples from Semi-Arid Countries: Rationale	30
Box 4.1	Water asset practices in Australia and Malta	60
Box 6.1	Water 4.0 case studies	88

1
Introduction

Abstract In this introductory chapter, climate change and the fourth industrial revolution—Industry 4.0—are briefly discussed as two strategic drivers that seem to be separate. Climate change is mentioned in the context of the required Level of Service (LOS) for water supply systems with water efficiency measures, seawater desalination, wastewater treatment with effluent reuse (or even water recycling). Built assets for water supply are closely linked to their natural catchments or natural assets. The concept of Industry 4.0 applied to the water sector, Water 4.0, can perhaps address Overall Equipment Effectiveness (OEE) and in parallel, many United Nations Sustainable Development Goals 6—Clean water and sanitation and 13—Climate action, in particular. The book examines, in a qualitative way, a proposition that Water 4.0 assists with the water supply decentralisation and sustainability, in particular climate resilience. Water 4.0 can be incorporated into the companies' Asset Management Systems (AMS) that should consider for this purpose a digital horizontal and vertical integration (*line of sight*) associated with Water 4.0. Such systems are built upon asset management fundamentals including adaptability. An alternative to adaptability remains a more

R. Kijak, *Water Asset Management in Times of Climate Change and Digital Transformation*, Palgrave Studies in Climate Resilient Societies,
https://doi.org/10.1007/978-3-030-79360-9_1

traditional approach, continuous improvement. Every organisation could set their own Water 4.0 related standards and objectives in their AMS and considers the preferred level of Water 4.0 adaptation. The Water 4.0 adaptation could arguably assist also with climate adaptation, and perhaps with some climate resilience. In this chapter, a brief overview of the remaining chapters has also been included.

Keywords Climate change · Industry 4.0 · Level of Service · Seawater desalination · Effluent reuse · Water recycling · Overall Equipment Effectiveness · Sustainable Development Goals · Asset Management System · Climate resilience

1.1 Strategic Drivers

Both climate change and the fourth industrial revolution (referred to further as Industry 4.0 or digital transformation) have been driving changes in water asset management practices in the recent years. They seem to be two separate drivers, but—as demonstrated throughout this book—digital transformation occurring within water utility companies, contributes clearly to climate resilience.

With weather anomalies breaking records in the recent years, climate change has become an evident risk to water assets operated by such companies. To ensure the required Level of Service (LOS), management of the water supply system involves presently maximising supply along with reducing demand by water efficiency measures or even water supply restrictions when necessary. In addition to seawater desalination, wastewater treatment with effluent reuse (or even water recycling) is increasingly considered as an integral part of the water supply system in circumstances when the natural sources fail to ensure the sufficient supply. This can involve recycled water named, for example, as Purified Recycled Water (PRW) and *new water* meeting high water quality standards. (See Glossary for definitions of terms and acronyms used throughout.)

Built water assets are closely linked to their natural catchments or natural assets. Managing them in parallel, in an integrated and systemic way, might assist in protecting the source water (natural supply). The catchments' meteorological and hydrological conditions can have direct impacts on both water quantity (e.g., hydrological drought) and quality (e.g., sediment laden runoff). Therefore, it is important that the adopted management practices within local catchments (or even regional/pan-regional practices) do not exacerbate the adverse impacts as with ash fallout from wild and bushfires. Relevant management practices for natural assets are then briefly considered in this book and supplement information for built assets.

Most of the identified examples of weather anomalies and their consequences (*lessons learned*) in this book originate from subtropical part of Australia (state of Queensland, in particular) where water utility companies have been struggling with evident climate impacts for, at least, the last 20 years. The recent extreme weather events in Australia have demonstrated water supply resilience. Despite occasionally introduced water use restrictions, all these extreme weather events have not generally interrupted water supply to the consumers so far. Other examples of managing assets for water supply originate from southern Europe (Malta) where water scarcity has always been a fact of life (Sapiano, 2018), but has been further exacerbated by progressing climate changes. Both countries have similarities, but they are also very different including, of course, their sizes and population density.

Another strategic driver in the water sector seems to be digital transformation associated with Industry 4.0—a new term coined at the Hannover Messe in 2011 and the subsequent 2016 World Economic Forum's meeting in Davos (e.g., Pfeiffer, 2017). The coronavirus pandemic is likely to expedite the digital transformation (e.g., Gulati, 2021). Industry 4.0's technologies blur the lines between engineering and science, and between various scientific disciplines. Operational technologies (OT) merge with information technologies (IT) (e.g., Gulati, 2021). The respective technologies include nine pillars: (i) *big data* and analytics; (ii) autonomous robots; (iii) simulation/*digital twins*; (iv) IoT/wireless sensors; (v) augmented/virtual reality; (vi) additive manufacturing; (vii) cloud computing; (vii) cybersecurity; and (ix) horizontal and vertical system integration (Alabi et al., 2019).

As Industry 4.0 reflects the fourth industrial revolution, Water 4.0 reflects the *fourth water revolution* (e.g., GWP, 2019). For this book, Water 4.0 is defined as a concept of Industry 4.0 applied to the water sector. Relevant technologies include most of the above Industry 4.0 pillar technologies. Water 4.0 can also result in the water sector decentralisation as a consequence of digitalisation (Alabi et al., 2019).

To implement the Water 4.0 concept, water utility companies need not only collect more data, but more importantly have proper analytical tools in place to convert data into information supporting optimal decisions. With Machine Learning, historical data might not be even required (e.g., Newhart et al., 2019).

1.2 Digital Transformation Improving Effectiveness and Efficiency

Industry 4.0 has been planned to address the economic effectiveness and efficiency, by *doing more with less* (e.g., Pfeiffer, 2017). Reportedly, the concept originated during the 2009 global financial crisis and was not related to technical innovations and the protection of environment, but to the improvement of Overall Equipment Effectiveness (OEE) that supports a greater economic growth (ibid.). As with the past economic growth that caused the current climate change impacts (e.g., WEF, 2018), Industry 4.0 is then likely to continue to adversely impact on the climate (e.g., Pfeiffer, 2017; WEF, 2018). However, many authors such as James Lovelock who is the originator of the Gaia theory, a former NASA scientist and a Fellow of the Royal Society, remain optimistic. Lovelock (2019) predicts that Artificial Intelligence will protect the climate better than humans alone.

Water is a valuable natural resource and requires a sustainable approach that can be achieved by minimising losses by a greater efficiency and less *waste* (Alabi et al., 2019). The concept of Industry 4.0 applied to the water sector, Water 4.0, can perhaps address these issues and in parallel, many United Nations Sustainable Development Goals (SDGs) (UN, n.d.). *Can Water 4.0 address SDGs 6—Clean water and sanitation and 13—Climate action, in particular? Can the greater OEE and climate protection be attained in parallel through digital transformation?* Even without the futuristic predictions, the improved OEE and less *waste* might intuitively translate into lower carbon and water footprints. This could perhaps be demonstrated if a Life Cycle Assessment (LCA) study with its *from cradle to grave* approach were applied. As such quantitative study is outside the scope of this book, can it be demonstrated, at least, that Water 4.0 enhances sustainability including climate resilience of the water sector? Such proposition is being examined in this book only in a qualitative way.

1.3 Water 4.0 Adaptation and Climate Resilience Through Asset Management Systems

To warrant asset performance, robustness, and sufficient funding, the water sector is one of the first industries that have embraced and implemented ISO 55001 (ISO, 2014b) based Asset Management Systems (AMS). The implementation of ISO 55001 principles has even become a requirement by the regulator as they are considered as a tool that assists with enhancing climate resilience in a cost-effective way (Seqwater 2017b).

During the fourth industrial revolution, the water utility company's AMS should also consider for this purpose a digital horizontal and vertical integration (*line of sight*) associated with Water 4.0 within the company. The relevant asset management fundamentals: value, alignment, leadership, assurance (ISO, 2014a) support this process, but more importantly a proposed new fundamental: adaptability (see Hardwick et al., 2020).

In addition to adaptability, the implementation of Water 4.0 through AMS could also occur through a more traditional approach, continuous improvement. Adaptability seems to be more relevant for technical innovations to bring the greatest business value through Water 4.0 adaptation. However, Water 4.0 requires a digital maturity (GWP, 2019). Moreover, adaptability of a water utility requires an asset management maturity (ibid.). It might then be appropriate for every organisation to set their own Water 4.0 related standards and objectives in their AMS and consider the preferred level of adaptation. In parallel to Water 4.0 adaptation, this includes climate adaptation. The latter might be a pragmatic approach, as 100% climate resilience is rarely possible. For this book, climate adaptation is an integral part of climate resilience.

1.4 Content of This Book

Following a discussion of climate change-related risks (or further *climate risks*) in selected semi-arid countries (Chapter 3), this book explores examples of management practices in these countries, which involve primarily centralised and large-scale regional water supply facilities with their respective natural catchments (Chapter 4). For this process, examples from water stressed Australia and southern Europe (Malta) are considered as *lessons learned*. It is proposed that protection against climate risks is, however, becoming a less realistic and complex task as time passes. Due to the weather unpredictability, it has been suggested in this book that the current design and operational standards and codes should be replaced in the not-so-distant future by Machine Learning algorithms. They are non-linear and non-stationary and thus aligned more closely with the unpredictability of real world. The unpredictability, or random nature, does not only apply to the weather, but also, in many cases, to asset health and asset-related processes.

The refocus on accommodating, retreating and avoiding risks is then becoming the preferred approach. In this context, a proposition is being examined that Water 4.0 does not only increase the sector's economic effectiveness and efficiency, but also assists with the water supply decentralisation and sustainability, in particular climate resilience (Chapters 5 and 6). Such examination is being conducted through the literature review and case studies.

Finally, the book considers the process of developing an AMS that might assist with climate resilience and adaptation, and captures for this process Water 4.0 technologies and solutions (Chapter 7). This chapter (7) contains, amongst other things, the author's reflections on the AMS scope, implementation and opportunities, but also explores contextually the recent standards on information management, alignment of financial and non-financial functions and cybersecurity.

References

Alabi, M. O., Telukdarie, A., & Janse van Rensburg, N. (2019, December). *Water 4.0: An integrated business model from an industry 4.0 approach.* Proceedings of the 2019 IEEE International Conference on Industrial Engineering and Engineering Management (IEEM). https://www.researchgate.net/publication/339021634_Water_40_An_Integrated_Business_Model_from_an_Industry_40_Approach. Accessed 8 July 2020.

GWP. (2019). *Water 4.0. made in Germany.* German Water Partnership.

Gulati, R. (2021). *Maintenance and reliability best practices* (3rd ed.). Industrial Press.

Hardwick, J., Killeen, M., Kohler, P., Lafraia, J., & Nugent, S. (2020, April 2). *Living asset management maturity*. Living Asset Management Think Tank Incorporated.

ISO. (2014a). *ISO 55000:2014 Asset management—Overview, principles and terminology.* Geneva, Switzerland: International Organization for Standardization.

ISO. (2014b). *ISO 55001:2014 Asset management—Management systems—Requirements.* Geneva, Switzerland: International Organization for Standardization.

Lovelock, J. (2019, July 4). *Novacene: The coming age of hyperintelligence* (1st ed). Penguin Books.

Newhart, K. B., Holloway, R. W., Hering, A. S., & Cath, T. Y. A. (2019, March 21). Data-driven performance analyses of wastewater treatment plants: A review. *Water Research, 157*(2019), 498–513. Elsevier B.V.

Pfeiffer, S. (2017, January 25). The vision of "Industrie 4.0" in the making—A case of future told, tamed, and traded. *Nanoethics,* 11(2017), 107–121. https://link.springer.com/article/10.1007/s11569-016-0280-3. Accessed 19 August 2020. Springer Nature B.V.

Sapiano, M. (2018, August 30). *Interview - Malta: Water scarcity is a fact of life.* Copenhagen, Denmark: European Environment Agency. https://www.eea.europa.eu/signals/signals-2018-content-list/articles/interview-2014-malta-water-scarcity. Accessed 3 February 2020.

Seqwater. (2017b, July 31). *2018 bulk water price review. Seqwater submission. Part A*. Seqwater.

UN. (n.d.). https://www.un.org/sustainabledevelopment/sustainable-development-goals/. United Nations. Accessed 19 August 2020.

WEF. (2018, September). *Harnessing the fourth industrial revolution for water*. Fourth Industrial Revolution for the Earth Series. World Economic Forum. https://www.weforum.org/reports/harnessing-the-fourth-industrial-revolution-for-water. Accessed 19 June 2020.

2

Glossary of Major Terms and Acronyms

Abstract This chapter defines all major terms and acronyms used throughout this book. They refer to several distinctive issues and disciplines such as asset management, climate change, water management and digital transformation. In most cases, their definitions originate from relevant international standards and other publications by recognised bodies and agencies.

Keyword All terms listed in this chapter

Adaptability	Adaptability is an ability to change as a consequence of changes occurring in the surrounding environment (Hardwick et al., 2020)

(continued)

R. Kijak, *Water Asset Management in Times of Climate Change and Digital Transformation*, Palgrave Studies in Climate Resilient Societies, https://doi.org/10.1007/978-3-030-79360-9_2

(continued)

Artificial Intelligence (AI)	AI is a generic term that studies human mind and intelligent systems (machines) that perform cognitive functions associated with human mind (e.g., Freeman, 2020; McKinsey, 2020). AI is a multidisciplinary science and includes mathematics, psychology, physics, biology, neuroscience and computer science (Freeman, 2020). It can be implemented through Machine Learning (ML) and Natural Language Processing (NLP) further subdivided into Neural Networks (NN), Deep Learning (DL), etc. (ibid.). See the ML definition below
Asset	Asset means "something that has potential or actual value to an organization", as defined by the International Organization for Standardization (ISO) standard 55000:2014 *Asset management—Overview, principles and terminology* (ISO, 2014a). This is a broad definition and may include, for example, financial assets and intellectual property that are clearly beyond the scope of this book
Asset management	Asset management, as defined in ISO 55000 (ibid.), means "coordinated activity of an organisation to realise value from assets". The organisation definition can include a public entity such as a water utility company or a local council
Asset Management Plan (AMP)	AMP, as defined in ISO 55000 (ibid.), means "documented information that specifies the activities, resources and timescales required for an individual asset, or a grouping of assets, to achieve the organisation's asset management objectives"

(continued)

(continued)

Asset Management System (AMS)	AMS, as defined in ISO 55000 (ibid.), is "a management system for asset management whose function is to establish the asset management policy and asset management objectives"
Strategic Asset Management Plan (SAMP)	SAMP, as defined in ISO 55000 (ISO, 2014a), means "documented information that specifies how organisational objectives are to be converted into asset management objectives, the approach for developing Asset Management Plans, and the role of the Asset Management System in supporting achievement of the asset management objectives". As explained by ISO 55002 (2018b), SAMP can also be referred to as an asset management strategy *The **definition of asset, asset management, Asset Management Plan, Asset Management System and Strategic Asset Management Plan**taken from **ISO 55000:2014 Asset management — Overview, principles and terminology**, is reproduced with the permission of the International Organization for Standardization, ISO. This standard can be obtained from any ISO member and from the website of the ISO Central Secretariat at the following address:* www.iso.org. *Copyright remains with ISO*
Built assets	Built assets mean, in summary, buildings and infrastructure (BSI, 2015)
Climate adaptation	Climate adaptation means "the process of adjustment to actual or expected climate and its effects. In human systems, adaptation seeks to moderate or avoid harm or exploit beneficial opportunities. In some natural systems, human intervention may facilitate adjustment to expected climate and its effects" (IPCC, 2014)

(continued)

(continued)

Climate resilience	Climate resilience means "the capacity of social, economic, and environmental systems to cope with a hazardous event or trend or disturbance, responding or reorganising in ways that maintain their essential function, identity, and structure, while also maintaining the capacity for adaptation, learning, and transformation" (IPCC, 2014). When a reference is made in this book to climate resilience, it may also include climate adaptation
Combined Sewer Overflow (CSO)	CSO means an overflow from a sewer accepting both wastewater and stormwater runoff due to a rainfall event of an intensity that the sewer has not been designed for
Continuous improvement	ISO 9001 defines continuous improvement as a "recurring activity to enhance performance" (ISO, 2015). *The **definition of continuous improvement**taken from **ISO 9000:2015: Quality management systems — Fundamentals and vocabulary**, is reproduced with the permission of the International Organization for Standardization, ISO. This standard can be obtained from any ISO member and from the website of the ISO Central Secretariat at the following address:* www.iso.org. *Copyright remains with ISO*
Cyber-Physical System (CPS)	CPS is a system that connects the digital and the physical worlds and is used as a human–machine interface (Poljak, 2018). It is a combination of physical processes with computation and networking embedded with sensors, actuators and processors (Alabi et al., 2019). Cyber-Physical Water Systems (CPWS) are used contextually (GWP, 2019). In this book, CPS and CPWS are used interchangeably

(continued)

(continued)

Digital twins	*Digital twins* are a form of CPS. *Digital twins* enable the analysis, optimisation and improvement of a process, product or service (e.g., Smart Water Magazine, 2020). *Digital twins* ensure more effective asset design, project execution and asset O&M by dynamically integrating data and information throughout the asset lifecycle (IET and Atkins). For this process, information originated from a *digital twin* is compared with information from a system (twin) under control The simplest *digital twin* referred to in this book includes and a model developed with BIM (Building Information Modelling) and a hydraulic model supplemented with real time monitoring data. The provided examples of a *digital twin* include a digital replica of physical assets such as a sewerage system or a Wastewater Treatment Plant (WWTP) where the digital replica and physical asset are mutually interconnected

(continued)

(continued)

Drought	As defined by Australian Bureau of Meteorology (BOM, n.d): "Drought is a prolonged, abnormally dry period when the amount of available water is insufficient to meet our normal use. (...) Because people use water in so many different ways, there is no universal definition of drought. It is measured in different ways and at different timescales: • Meteorologists monitor the extent and severity of drought in terms of rainfall deficiencies (or shortages, compared to average rainfall for the period) • Agriculturalists rate the impact on primary industries • Hydrologists examine surface and groundwater levels • Sociologists define it by social expectations and perceptions and the impact on the community" Reproduced by permission of Bureau of Meteorology, © 2020 Commonwealth of Australia
Economic Level of Leakage (ELL)	"The point at which the cost of reducing leakage is equal to the benefit gained from further leakage reductions" (Ofwat, 2007) Ofwat does not consider that the sustainable ELL, as currently measured, is suitable to set leakage performance levels for the 2020–2025 period, see Ofwat (2020)

(continued)

(continued)

El Niño-Southern Oscillation (ENSO)	ENSO cycle includes opposite phases El Niño and La Niña and describes (NOAA, 2020a): "the fluctuations in temperature between the ocean and atmosphere in the east-central Equatorial Pacific (approximately between the International Date Line and 120 degrees West). (...) El Niño and La Niña episodes typically last nine to 12 months, but some prolonged events may last for years. While their frequency can be quite irregular, El Niño and La Niña events occur on average every two to seven years. Typically, El Niño occurs more frequently than La Niña. (...) The term El Niño refers to the large-scale ocean–atmosphere climate interaction linked to a periodic warming in sea surface temperatures across the central and east-central Equatorial Pacific. (...) The presence of El Niño can significantly influence weather patterns, ocean conditions, and marine fisheries across large portions of the globe for an extended period"
Exposure	Exposure means "the presence of people, livelihoods, species or ecosystems, environmental functions, services, and resources, infrastructure, or economic, social, or cultural assets in places and settings that could be adversely affected" (IPCC, 2014)

(continued)

(continued)

Hazard Analysis and Critical Control Points (HACCP)	HACCP is an internationally recognised system used to identify and manage risk to food products and might be applied to drinking water. Australian Drinking Water Guidelines (ADWG) developed by the National Health and Medical Research Council of Australia (NHMRC, 2018) recommends the development of, amongst other things, a HACCP system that uses ADWG as its framework. As with ISO 22000 (ISO 2018a), the HACCP implementation is compatible with the implementation of a quality management system in accordance with ISO 9001 (ISO, 2015) or an Asset Management System in accordance with ISO 55001 (ISO, 2014a). Unlike HACCP, ISO 22000 is broader and focused on the whole supply chain, for example, *source-to-tap* (or even *catchment-to-tap*) and *toilet-to-tap*
Hazard	Hazard means "the potential occurrence of a natural or human-induced physical event or trend or physical impact that may cause loss of life, injury, or other health impacts, as well as damage and loss to property, infrastructure, livelihoods, service provision, ecosystems, and environmental resources" (IPCC, 2014)

(continued)

(continued)

Hybrid analytics	For the purpose of this book, hybrid analytics involves real time data analyses without ML assistance to statistically identify anomalies and predict the potential asset or process failure. As with the ML based predictive analytics, it is a probabilistic tool used to predict what might happen In this instance, the operator enters periodically patterns of anomalies in the operational software solution, which have been identified based on historical data. When anomalies have been detected, the solution and/or the operator conducts the subsequent cause-effect analysis to identify asset failure modes or process adjustments required Unlike ML controlled predictive analytics, the operator can only focus on the relationship between one or two pairs of features in the anomaly patterns (Reddy, 2020). Hybrid analytics does not capture non-linear and non-stationary processes When a general reference is made to predictive analytics in this book, this term might include both predictive analytics (ML controlled) and hybrid analytics (operator controlled)
Infrastructure Leakage Index (ILI)	ILI means "the ratio between actual real losses and an estimate of the minimum real losses that could be technically achieved for the system operating pressure, average service connection length and service connection density" (Alegre et al., 2006)

(continued)

(continued)

Internet of Things (IoT)/Industrial Internet of Things (IIoT)	IoT/IIoT is a rapidly growing network of sensors and actuators embedded with software and network connectivity, which can collect and exchange data over the internet and assist with real time and automated solutions (e.g., Stankovic et al., 2020; WEF, 2018). GWP (2019) refers to Internet of Things and Services (IoTS) instead
Level of Service (LOS)	LOS is a set of parameters in the process of delivery of services to the customers and other stakeholders, which are focused on social, environmental, economic and other outcomes with the parallel realisation of value from assets (assetinsights.net, n.d.; ISO, 2014b). In the water supply context, LOS objectives refer to the security for both quantity and quality of water to be supplied to the customers (e.g., Seqwater, 2017a)
Life Cycle Assessment (LCA)	LCA means a "compilation and evaluation of inputs and the potential environmental impacts of a product throughout its life cycle" (ISO, 2006) *"The **definition of Life Cycle Assessment (LCA)**taken from **ISO 14040:2006: Environmental management— Life cycle assessment — Principles and framework**, is reproduced with the permission of the International Organization for Standardization, ISO. This standard can be obtained from any ISO member and from the website of the ISO Central Secretariat at the following address:* www.iso.org. *Copyright remains with ISO."*

(continued)

(continued)

Machine Learning (ML)	ML is a form of AI, which learns from data through training (Freeman, 2020). The respective model used has parameters and weights that are being adjusted, as the model learns different algorithms (ibid.). Real time and/or historical data are required for the algorithms to be trained or tested (ibid.) Neural networks (NN)—often considered as part of ML—is the most frequently used method for water and wastewater systems (e.g., Hadjimichael et al., 2016). NN classifies data or finds the complex relationship between variables in regression analyses, for example (McKinsey, 2020). These relationships can be non-linear and non-stationary to reflect the random and unpredictable nature of subject events. They do not require a prior knowledge of the underlying mechanism and historical data (Newhart et al., 2019). This assists with the identification of asset health and process anomalies and their subsequent classification, that is, a cause-effect relationship (e.g., Reddy, 2020) In this book, AI and ML might be used interchangeably. See also the definition of AI above
Natural assets	Natural assets include, in summary, flora, fauna, land, water, subsoil and air (UN, 1997)
Non-Revenue Water (NRW)	NRW means water that has been produced at a water treatment plant, but subsequently unaccounted for because of pipe leaks and bursts, water thefts, etc. (i.e., water losses)

(continued)

(continued)

Overall Equipment Effectiveness (OEE)	OEE is a metric for measuring improvement in machine reliability and effectiveness. It can be estimated by multiplying machine *availability*, *efficiency* and *quality performance* by taking into account respective losses (e.g., Defeo, 2017)
PLC	PLC means Programmable Logic Controller
Purified Recycled Water (PRW)	PRW means a WWTP effluent treated further to the drinking water and/or irrigation standards, which is also known as *new water* or *NEWater* (The latter term was coined in Singapore in 2003, see Brears, 2017). In Malta, a name Highly Polished Reclaimed Water (HPRW) is also used in parallel with *new water* (e.g., Water Services Corporation, n.d.)
Riverine flooding	Riverine flooding means flooding from overflows of rivers and creeks following a rainfall event occurring in a catchment (JBA Group, 2019)
SCADA	SCADA means Supervisory Control and Data Acquisition
Total Expenditures (TotEx)	TotEx is a sum of CapEx (Capital Expenditures) and OpEx (Operational Expenditures) over a defined period, as defined by ISO 55010 (ISO, 2019)
Virtual sensors	Virtual sensors can predict by using AI the output of physical sensors if they were installed at given locations. Therefore, they can supplement or even replace physical sensors with improved efficiency and lower cost (e.g., WEF, 2018)
Vulnerability	Vulnerability means "the propensity or predisposition to be adversely affected. Vulnerability encompasses a variety of concepts and elements including sensitivity or susceptibility to harm and lack of capacity to cope and adapt" (IPCC, 2014)

(continued)

(continued)

Waste	*Waste* is defined in *Lean* as any activity consuming resources, but bringing no value to the customer (e.g., Defeo, 2017)
Water 4.0	According to GWP (2019), "Water 4.0 focuses on digitalization and automation as the core aspects of a strategy for resource-efficient, flexible and competitive water management In this regard, and in analogy to the Industry 4.0 initiative, Water 4.0 refers to core characteristics and concepts of this industrial revolution such as the networking of machines, processes, storage systems and operational resources, along with smart grids and the Internet of Things and Services As an umbrella term, Water 4.0 brings these elements into a systematic relationship within the water management context In the implementation of Water 4.0, Cyber-Physical Systems (CPS) produce an optimal level of networking between virtual and real water systems, in which software tools are used throughout the planning, construction and operations phases. This will enable the creation of an intelligent network that links water users (i.e., agriculture, industry and households) and components within a sustainable water infrastructure, while also drawing on data from the environment and the water cycle, allowing for a holistic approach along the entire value chain In addition, Water 4.0 provides a high level of transparency for water users, thus covering current needs, while also providing opportunities for creative, future focused jobs in the water sector."

(continued)

(continued)

Water supply asset	Water supply asset, for the purpose of this book, means both built and natural assets such as water storages (e.g., dams and groundwater aquifers) used for raw water abstraction, their catchments including groundwater recharge zones, Water Treatment Plants (WTPs), desalination plants, water transmission pipelines, water distribution networks and WWTPs if their effluent is further treated by non-conventional processes to become, for example, Purified Recycled Water (PRW) in Queensland or *new water* in Malta Robust multiple barriers appropriate to the level of potential contamination and the effluent reuse option are required in accordance with relevant guidelines (e.g., NRMMC, EPHC & AHMC, 2006). Both PRW and *new water* should be meeting the drinking water standard before the next barrier (that is, natural environment—by replenishing groundwater and surface water storages)

References

Alabi, M. O., Telukdarie, A., & Janse van Rensburg, N. (2019, December). *Water 4.0: An integrated business model from an industry 4.0 Approach.* In the proceedings of the 2019 IEEE International Conference on Industrial Engineering and Engineering Management (IEEM). https://www.researchgate.net/publication/339021634_Water_40_An_Integrated_Business_Model_from_an_Industry_40_Approach. Accessed 8 July 2020.

Alegre, H., Baptista, J. M., Cabrera Jr., E., Duarte, P., Merkel, W., & Parena, R. (2006). *Performance indicators for water supply services* (2nd ed., Table 21, p. 30). IWA Publishing. © IWA Publishing.

assetinsights.net. (n.d.). https://assetinsights.net/Glossary/G_Level_of_Service.html. Accessed 6 December 2020.

Brears, R. (2017, January 17). Singapore transitioning towards urban water security. In R. Brears (Ed.), *Urban Water Security* (1st ed., pp. 225–241). Wiley. https://doi.org/10.1002/9781119131755.ch13. Accessed 10 March 2020.

BSI. (2015). *Publicly Available Specification (PAS) 1192–5:2015 specification for security-minded building information modelling, digital built environments and smart asset management.* British Standards Institution.

Defeo, J. A. (2017). *Juran's quality handbook* (7th ed.). McGraw-Hill Education.

Freeman, R. (2020, July 24). When to NOT use AI or use it based on my experience. *Towards Data Science online*. https://towardsdatascience.com/when-to-not-use-ai-or-use-it-based-on-my-experience-abb58c063aba. Accessed 8 July 2020.

GWP. (2019). *Water 4.0. Made in Germany.* German Water Partnership.

Hadjimichael, A., Comas, J., & Corominas, L. (2016, December 1). *Do artificial intelligence methods enhance the potential of decision support systems? A review for the urban water sector.* To appear in: AI Communications.

Hardwick, J., Killeen, M., Kohler, P., Lafraia, J., & Nugent, S. (2020, April 2). *Living Asset Management Maturity*. Living Asset Management Think Tank Incorporated.

IPCC. (2014). *Climate change 2014: Impacts, adaptation, and vulnerability. Part A: Global and sectoral aspects. Contribution of working Group II to the Fifth Assessment Report of the Intergovernmental Panel on Climate Change* (C. B. Field, V. R. Barros, D. J. Dokken, K. J. Mach, M. D. Mastrandrea, T. E. Bilir, M. Chatterjee, K. L. Ebi, Y. O. Estrada, R. C. Genova, B. Girma, E. S. Kissel, A. N. Levy, S. MacCracken, P. R. Mastrandrea, & L. L. White, Eds., pp. 1–1131). Cambridge University Press.

ISO. (2006). *ISO 14040:2006. Environmental management—Life cycle assessment—Principles and framework.* Geneva, Switzerland: International Organization for Standardization.

ISO. (2014a). *ISO 55000:2014 Asset management—Overview, principles and terminology.* Geneva, Switzerland: International Organization for Standardization.

ISO. (2014b). *ISO 55001:2014 Asset management—Management systems—Requirements.* International Geneva, Switzerland: International Organization for Standardization.

ISO. (2015). *ISO 9001:2015 Quality management systems—Quality management systems—Requirements.* Geneva, Switzerland: International Organization for Standardization.

ISO. (2018a). *ISO 22000:2018 Food safety management systems—Requirements for any organization in the food chain.* Geneva, Switzerland: International Organization for Standardization.

ISO. (2018b). *ISO 31000:2018 Risk management—Guidelines.* Geneva, Switzerland: International Organization for Standardization.

ISO. (2019). *ISO/TS 55010:2019 Asset management—Guidance on the alignment of financial and non-financial functions and asset management.* Geneva, Switzerland: International Organization for Standardization.

JBA Group. (2019, March 7). *One month on: A retrospective look at the Townsville flooding in February 2019.* https://www.jbarisk.com/flood-services/event-response/a-retrospective-of-townsville-flooding/. Accessed 23 January 2020.

McKinsey. (2020). *An executive's guide to AI*. McKinsey & Company. https://www.mckinsey.com/business-functions/mckinsey-analytics/our-insights/an-executives-guide-to-ai. Accessed 8 December 2020.

Newhart, K. B., Holloway, R. W., Hering, A. S., & Cath, T. Y. A. (2019, March 21). Data-driven performance analyses of wastewater treatment plants: A review. *Water Research, 157*, 498–513. Amsterdam, The Netherlands: Elsevier B.V.

NHMRC. (2018). *National water quality management strategy. Australian drinking water guidelines 6 2011.* Version 3.5 updated on August 2018. Canberra, Australia: National Health and Medical Research Council.

NRMMC, EPHC, & AHMC. (2006). *National water quality management strategy. Australian guidelines 21 for water recycling. Managing health and environmental risks.* Canberra, Australia: Natural Resource Management Ministerial Council, Environment Protection and Heritage Council, Australian Health Ministers Conference.

NOAA. (2020a). *What are El Niño and La Niña?* Washington, DC, United States: National Ocean Service. Last updated on 10 February 2020. https://oceanservice.noaa.gov/facts/ninonina.html. Accessed 12 February 2020.

Ofwat. (2007, November 13). *Providing best practice guidance on the inclusion of externalities in the ELL calculation, Main Report v05.* Ref: PROC/01/0075,

Birmingham, UK: Office of Water Services. Licensed under the Open Government Licence.

Ofwat. (2020, May). *Reference of the PR19 final determinations: Cost efficiency – response to common issues in companies' statements of case* (p. 56). Birmingham, UK: Office of Water Services. Licensed under the Open Government Licence.

Poljak, D. (2018). *Industry 4.0 – New challenges for public water supply organizations.* https://bib.irb.hr/datoteka/940425.INDUSTRY_4.0__New_Challenges_for_Public_Water_Supply_Organizations.pdf. Accessed 8 July 2020.

Reddy, R. (2020, June 17). Machine learning 101 in predictive maintenance. *IndustryWeek*. https://www.industryweek.com/technology-and-iiot/article/21134278/machine-learning-101-in-predictive-maintenance. Accessed 8 July 2020.

Seqwater. (2017a, March). *Queensland bulk water supply authority, trading as Seqwater.: Water for life South East Queensland's Water Security Program 2016–2046*. Version 2. Ipswich, Australia: Seqwater.

Smart Water Magazine. (2020). The digital twin, another step forward in ACCIONA's commitment to technology. *Smart Water Magazine*, SWM Monthly 3. https://smartwatermagazine.com/blogs/julio-de-la-rosa/digital-twin-another-step-forward-accionas-commitment-technology. Accessed 7 August 2021.

Stankovic, M., Hasanbeigi, A., Neftenov, N., & Tambourine Innovation Ventures. (2020, April). *Use of 4IR technologies in water and sanitation in Latin America and the Caribbean.* Washington, DC, United States: Inter-American Development Bank, Water and Sanitation Division.

UN. (1997). *Glossary of environment statistics.* Studies in Methods, Series F, No. 67, New York, NY, Unites States: Department for Economic and Social Information and Policy Analysis, Statistics Division, United Nations.

Water Services Corporation. (n.d.). *New water.* http://www.wsc.com.mt/information/new-water/. Accessed 15 January 2020.

WEF. (2018, September). *Harnessing the Fourth Industrial Revolution for Water.* Fourth Industrial Revolution for the Earth Series. Geneva, Switzerland: World Economic Forum. https://www.weforum.org/reports/harnessing-the-fourth-industrial-revolution-for-water. Accessed 19 June 2020.

3

How Can Climate Change Impact upon Water Supply Assets?

Abstract An iterative risk management approach to climate change, which considers vulnerability, hazards, exposure (risk assessment) and adaptation and resilience (risk treatment), is discussed in this chapter. Climate change impact projections with general resilience/adaptation measures related to water supply assets and water resources for Australia and southern Europe are outlined in this chapter. They are followed by a summary of 2019–2020 weather conditions for Australia and Malta (southern Europe). Subtropical Australia and Malta have been used as contextual examples in this book. Weather anomalies and extreme weather events such as Australia's recent drought and megafires (2019–2020) and the earlier Millennium Drought (1999–2010), Queensland Flood (2011) and a flood in Townsville in the Northern Queensland (2019) are discussed. Subsequently, these events' impacts on Level of Service (water supply services) are explored in more detail. They include impacts on water quantity and quality and hence on safety, quality, reliability and availability of supply, environmental sustainability and the last, but not least, cost.

R. Kijak, *Water Asset Management in Times of Climate Change and Digital Transformation*, Palgrave Studies in Climate Resilient Societies,
https://doi.org/10.1007/978-3-030-79360-9_3

Keywords Risk management · Risk resilience · Risk adaptation · Climate change projections · Weather anomalies · Millennium Drought · Queensland Flood · Level of Service

Climate change is clearly an asset risk. Though the average global temperature has only risen slightly above 1 °C above the preindustrial 1850–1900 average (C3S, 2021), climate change seems to be responsible already for extreme weather events such as floods and droughts with greater frequencies and intensities or simply temperature extremes (heatwaves) and erratic rainfall (ibid.). However, most of climate change impacts will occur in a relatively distant future what contributes to their uncertainty (risk), particularly at the local scale.

3.1 A Risk-Based Approach to Climate Change

3.1.1 Risks to Water Utilities

Clearly, there is a relationship between climate change, water crises and water utility companies' asset management processes and practices. Over the last several years, both climate change-related impacts and water crises have been identified by World Economic Forum (WEF) as the top five greatest risks (WEF, 2020). Asset management is generally considered as the most important business process carried out by water utility companies to ensure uninterrupted water supply (e.g., Gourbesville, 2019). Good asset management can reportedly contribute to seven Sustainable Development Goals (SDGs) including SGD 6—Clean water and sanitation and SGD 13—Climate action (ISO, 2018b).

3.1.2 Risk Assessment and Treatment

Risk management is commonly applied to deal with uncertainty and potential impacts (hazards) as risk is by definition an "effect of uncertainty on objectives" in accordance with the ISO 31000 standard on risk management (ISO, 2018a). An iterative risk management approach is applied in this book, which considers exposure, vulnerability, resilience and adaptation (e.g., IPCC, 2014), as depicted in Fig. 3.1. This is broadly consistent with the ISO 31000 risk management standard, but instead of consequence and likelihood, a disaster risk approach has been applied with vulnerability and exposure coupled with hazards (IPCC, 2014 and Risk Society).

Risk treatment follows such risk assessment with measures that consider (ibid.): (i) climate variables that make the system vulnerable

Fig. 3.1 An iterative risk management approach in the climate change context

(e.g., extreme precipitation or a prolonged drought); (ii) environmental variables (e.g., a bed and banks of a river or creek), but such projections of the future changes are highly uncertain. *Climate resilience requires moving the goalposts with climate changes progressing.* New hazards, exposure and vulnerability require new measures to increase resilience. In this book, climate resilience may include adaptation.

Relevant definitions of climate resilience and adaptation have been included in Glossary. From the psychological viewpoint, adaptation means an ability of an individual to preserve the existing resources (Wong-Parodi et al., 2015). Resilience means a general ability to recover and even emerge stronger following exposure to stressors/hazards (ibid.)

3.1.3 Examples of Climate Risks and Risk Management Practices

In this chapter and Chapter 4, climate risks are discussed based on examples from semi-arid and water stressed countries with the respective rationale provided in Box 3.1.

Box 3.1 Examples from Semi-Arid Countries: Rationale

Contextual examples from water stressed Australia and southern Europe are considered in this chapter and Chapter 4 *lessons learned.* They involve a subtropical part of Australia (particularly, subtropical Queensland including Townsville) and Malta in the middle of Mediterranean Sea (southern Europe). The examples may originate from the subtropical region, but most issues raised are perhaps applicable also to the temperate region. The subtropical region has already *learned lessons*

Australia and Malta have similarities, but they are also very different including, of course, their sizes and population density with Malta's staggering 1400 persons per square kilometre (Sapiano, 2018). They seem to be the most climate vulnerable regions amongst the developed countries.

As with Australia, Malta is one of the driest countries in the world with the lowest natural water resource per capita amongst the Mediterranean countries (Gourbesville, 2019). They have recently been experiencing hotter summers, temperature variability/extremes and generally drier conditions (see Tables 3.1 and 3.2). In both countries, seawater desalination has become an established technology and Purified Recycled Water (PRW) has also found some community acceptance.

There are also differences. Unlike Australians who only recently have learned how to use water more efficiently, Maltese have been using merely 110 L per day per person with plans to reduce it even further (ibid.). For coastal parts of Australia, long-term droughts are examples of extreme weather anomalies, but for Malta, water scarcity has always been a fact of life (Sapiano, 2018). Riverine floods do not occur in Malta because there are no rivers and creeks present, but flash flooding has recently been occurring more often and with a greater intensity. Maltese Islands are generally relying on groundwater resources supplemented by seawater desalination, PRW and *stormwater harvesting* (Sapiano, 2018). In Australia, the only metropolitan area that relies on groundwater resources is Perth (Western Australia). The rising sea levels that threaten Malta's underground freshwater resource (Gourbesville, 2019) have a greater impact than in Australia due to the country's smaller size and greater population density.

Therefore, droughts, floods and fires in Australia discussed in this book have no equivalent events in Malta. The issue of natural catchment management is more relevant for Australia, as surface water is less used in Malta for water supply.

Table 3.1 contains summaries of climate change impact projections with the recently adopted resilience/adaptation measures related to water supply assets and water resources for Australia and southern Europe (*2014 perspective*), as outlined in the Intergovernmental Panel on Climate Change (IPCC) Fifth Assessment Report (unless otherwise stated).

The 2014 IPPC projections and identified hazards have generally been consistent with the actual weather conditions over the next several years. Globally, 2016 and 2020 were the warmest years on record that began in 1850 (CS3, 2021). The last decade (2010–2019) was the warmest on record (ibid). 2019 was the second warmest year on record (C3S, 2020). Table 3.2 contains a summary of weather conditions in 2019 and 2020 for Australia and Malta.

Table 3.1 Summaries of the projected climate change hazards and the adopted resilience/adaptation measures (2014 perspective) for Australia and southern Europe (Hewitson et al., 2014; Kovats et al., 2014; Reisinger et al., 2014, unless otherwise stated/updated in the table)

No	Projected climate change hazards		Resilience and adaptation measures	
1	Australia	Southern Europe	Australia	Southern Europe
2	• Decline in precipitation in southern Australia, particularly in southwestern Australia, but with an increase in most of Australia's regions in the intensity of daily (or shorter) rainfall extremes • Drought events to increase in southern Australia • Freshwater resources projected to decline in southeastern and southwestern Australia (inland) • Increased frequency and intensity of flood damage to settlements and infrastructure in Australia due to more intense extreme rainfall events driven by a warmer and wetter atmosphere	• A high variability in rainfall and runoff, as well as a water temperature increase • Water quality impacts due to an increased nutrient loading (increased evapotranspiration and increased nutrients concentrations) • Impeded economic activity in southern Europe more than in other sub-regions in Europe with multiple sectors adversely affected including farming and infrastructure	• Resilience/adaptation included in planning processes though often only at the conceptual level • Integrated responses encompassing: (i) management of demand through water conservation/efficiency across all sectors; and (ii) management of supply involving seawater desalination and water recycling • Since the early 2000s, water conservation measures have reduced Melbourne's total per capita water use by 40% and Brisbane's by about 50%. Adoption of water recycling and rainwater harvesting resulted in up to 60% water savings in some parts of Adelaide (South Australia)	• Adaptation policy developed at the European Union, national and local government levels for compliance with the Water Framework Directive and Flood Directive • Increased irrigation to offset the rainfall and runoff variability is unlikely to become an option (particularly, in the Mediterranean area) because of: (i) projected declines in total runoff and groundwater resources; (ii) decline in groundwater recharge; and (iii) increase in potential demand for such irrigation

No	Projected climate change hazards		Resilience and adaptation measures	
1	Australia	Southern Europe	Australia	Southern Europe
	• Constraints on water resources in southern Australia, driven by rising temperatures and reduced winter rainfall • Water-borne diseases projected to increase with increased cases of bacterial gastroenteritis, salmonellosis, zoonotic diseases such as cryptosporidiosis and giardiasis • Increased morbidity, mortality, and infrastructure damages during heatwaves in Australia, resulting from increased frequency and magnitude of extreme high temperatures; vulnerable populations include the elderly and those with existing chronic diseases; population increases and ageing trends constrain effectiveness of adaptation responses		• Policy and institutional reforms including the 2004 National Water Initiative and 2007 Commonwealth Water Act. The establishment of the National Water Commission in 2004 and the Murray-Darling Basin Authority in 2008 • Water recycling initiatives and established seawater desalination technology in place (see Box 2) • The flood risk adaptation assisted with a new edition of Australian Rainfall and Runoff (ARR) Guideline for design flood estimation (Ball et al., 2019). *ARR Guideline was finalised in 2019, that is after the issue of IPCC Fifth Assessment Report*	• The adverse effect of climate change on water resources in coastal regions of southern Europe might be further exacerbated by constructing desalination plants due to the associated high energy consumption (i.e., greenhouse gas emissions)

(continued)

Table 3.1 (continued)

No	Projected climate change hazards		Resilience and adaptation measures	
1	Australia	Southern Europe	Australia	Southern Europe
	• Increased damages to settlements and associated infrastructure from bushfires in most of southern Australia • A downward trend in water supply capability for water storages (lower inflows) to meet the future demand (Nguyen et al., 2020)		• Adaptation options in urban areas include ecosystem-based approaches such as retaining floodplains and floodways, restoring wetlands, and retrofitting existing systems to attenuate flows • Changes to the operations of water storages to minimise the likelihood of floods • Revised insurance practice to cover flood damages	

Table 3.2 A summary of weather conditions (rainfall and temperatures only) for Australia and Malta in 2019 and 2020 (BOM 2020a, 2021 respectively). Reproduced by permission of Bureau of Meteorology, © 2020 Commonwealth of Australia. Maltese Islands Weather 2020 and Maltese Islands Weather 2021 respectively

No	Year	Weather Conditions	
		Australia	Malta
1	2019	• The warmest year on record with the annual mean temperature 1.52 °C above average • Mean maximum and minimum temperatures 2.09 °C and 0.95 °C above average respectively • The driest year on record with the averaged rainfall 40% below average at 277.6 mm • Rainfall below average for most of Australia except for parts of Queensland's northwest and northern tropics • Much of Australia impacted by drought particularly severe in New South Wales and southern Queensland • Severe fire weather throughout the year • El Niño–Southern Oscillation neutral throughout the year	• The warmest year in the previous decade (2010–2019) with the annual mean exceeded by 1.3 °C with the maximum mean exceeded by 1.8 °C • All winter and spring months colder than average • All summer and autumn months warmer than average • 2019 total rainfall close to average (559.2 mm) • November and October receiving approximately half of the total annual rainfall • Both spring and autumn wetter than average • As usual, summer extremely dry
2	2020	• The fourth-warmest year on record with the mean temperature 1.15 °C above average • Mean maximum and minimum temperatures 1.24 °C and 1.05 °C above average respectively • The beginning of the year affected by drought, extreme heat and widespread bushfires in eastern Australia • Much of southeastern and eastern Australia affected by heatwaves in November	• The annual mean temperature exceeded by 1.6 °C with the maximum mean exceeded by 2.0 °C • June, July and October—the only months with temperatures not above the monthly minimum means • The annual rainfall total—200 mm below the average • The rainfall total of 381.7 mm including 112.1 mm recorded during a single day (14 September 2020)

(continued)

Table 3.2 (continued)

No	Year	Weather Conditions	
		Australia	Malta
		• Australia's averaged rainfall of 483.4 mm, that is 4% above average • Flooding in eastern Australia during February and March, particularly in Queenslanc • The El Niño–Southern Oscillation initially neutral and finally declared in September	• The second driest January on record • The driest February ever with only 0.5 mm of rain, that is as with the typical dry summer months • Only 57.5 mm of rain recorded in the first six months of the year

3.2 Flood-Drought Cycle

3.2.1 Drought

2019–2020 Drought and Megafires in Australia

In late 2019 and early 2020, hot and dry conditions coincided with extreme wildfire and bushfire events (*megafires*) nationwide (NOAA, 2020b). *Megafires* began in Queensland in September and subsequently progressed southward as a result of an extremely dry and warm spring and early summer (ibid.), see Fig. 3.2.

Megafires were soon followed by severe rainfall events particularly in southeast Queensland and northeastern New South Wales, but the above average rainfall in January 2020 was not sufficient to quickly clear long-term rainfall deficiencies. (Readers may refer to the relevant monthly weather update for complete information, BOM, 2020b.)

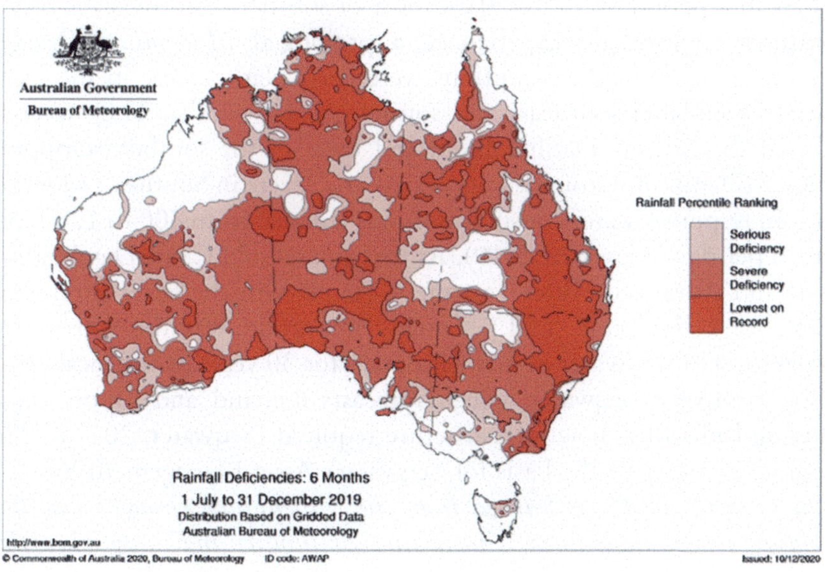

Fig. 3.2 Rainfall deficiencies for the period 1 July–31 December 2019 (BOM 2020b) (Reproduced by permission of Bureau of Meteorology, © 2020 Commonwealth of Australia)

Megafires burnt through 18.6 million ha of land with a total perimeter of 19,235 km (Roorda, 2020).

As a result of the prolonged drought and a general downward trend of the dams' accessible storage capacity in Australia (see Nguyen et al., 2020), media reported in early 2020 that a number of regional towns had run out of river water supplies (Water Source, 2020. Several local councils were required to cart water from other areas or required to use brackish groundwater that was treated by mobile desalination plants (ibid.).

Demand management and water restrictions seem to remain part of asset planning in Australia even if there are more plans for building regional large-scale seawater desalination plants. In fact, the projected lower inflows to water storages combined with the population growth are likely to lead to an increase in severity and frequency of water restrictions in Australia in the future.

1999–2010 *Millennium Drought*

Earlier in a period of 1996–2010, most of southern Australia including Southeast Queensland experienced a prolonged *Millennium Drought* (see Fig. 3.3). Drought conditions were particularly severe in the more densely populated southeast and southwest of Australia. They severely affected the Murray-Darling Basin and most of the southern cropping zones. Millennium Drought was particularly severe in Southeast Queensland. In response, water consumption was reduced from 300 to 120 L per person per day (Seqwater, 2019). At that time, Gold Coast Desalination Plant and Advanced Water Treatment Plants with the latter producing PRW, were also built (ibid.). Equally important was the subsequent preparation of the Water Security Program for 30 years by the local water utility company, Seqwater, which forecasts demand and supply capability and identifies new infrastructure required (Seqwater, 2017a). The program is presently the basis for Seqwater's Asset Management System.

In general, the lesson learned from the Millennium Drought was that building water supply assets is more time consuming and costly when they are required to be built quickly so it is important to plan earlier (Seqwater,

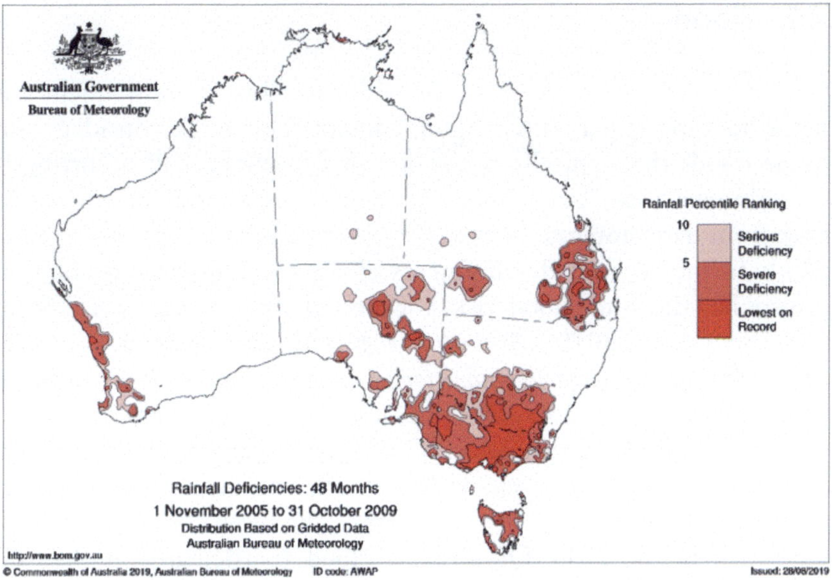

Fig. 3.3 Rainfall deficiencies for the period 1 November 2005–31 October 2009 (BOM, 2019) (Reproduced by permission of Bureau of Meteorology, © 2020 Commonwealth of Australia)

2019). For this process, sustainability and adaptability of measures outlined in asset plans should be considered due to: (i) uncertainty of climate change projections at the local level; (ii) future availability of new technologies; and (iii) changing community expectations. For example, the use of large scale and regional desalination plants might be perceived as less sustainable due to their high energy demand and adversely impacting on water efficiency measures, water recycling initiative and absorption of new technologies such as small solar powered desalination plants located closer to the consumers (e.g., Xu et al., 2020) and forward osmosis (FO) that needs no external energy (e.g., Rabiee et al., 2019). However, the perception of large-scale desalination plants as less sustainable might change if most of electricity supplied to the grid originated from renewable sources. *Therefore, a question that needs to be answered is adaptability and reversibility of climate resilience measures based on such risks (uncertainties)?*

3.2.2 Flood

After the drought comes the flood. As with droughts, more frequent and intense flooding is part of changing climate. This book considers only riverine floods that typically occur due to a heavy rainfall occurring in the catchment areas, but it might also occur as a result of emergency release from dam storages.

With further coastal developments along the Queensland coastline in Australia, the total flood risk is expected to increase by 130% by 2100 (JBA Group, 2019). *As with asset planning for droughts, it is necessary to plan all necessary upgrades for managing riverine floods earlier so that all water supply assets can withstand and recover from floods that become more extreme as time passes.* This is particularly so in Queensland where the drought-flood cycle with an increasing variability (frequency and intensity) has become the new norm. The recent devastating flood events include the 2011 Queensland Flood (particularly in Brisbane and Ipswich in Southeast Queensland) and the more recent flood in Townsville (North Queensland) in 2019, though it is debatable whether, in fact they were influenced by climate change. With respect to water supply, floods can impact on both water treatment and transmission and distribution by: (i) changing source water quality (and thereby reducing water treatment capacity); (ii) breaking water mains; and (iii) cutting power.

Queensland Flood

The key meteorological and hydrological conditions leading to the 2011 Queensland Flood (January 2011) have been summarised in Table 3.3.

Lessons learned from the 2011 Queensland Flood might fall into three categories:

- Documenting and adhering to procedures for the operations of dam storages (i.e., asset operations manuals or plans) by the operator
- Downstream land-use planning
- Water quality deterioration at the intake from contamination within catchments.

Table 3.3 The key meteorological and hydrological conditions leading to the 2011 Queensland Flood

No	2011 Queensland Flood—Key meteorological and hydrological conditions
1	The flood followed an extremely wet spring (September–November) with catchments already saturated (van den Honert & McAneney, 2011)
2	Total rainfall recorded by some stations west of Brisbane was approximately 1200 mm over December 2010 and January 2011 (ibid.)
3	The lower Brisbane River catchment (downstream of the dams) was experiencing significant rainfall, for example, the total of approximately 450 mm in the Lockyer Creek catchment (ibid.). The rainfall intensity exceeded an average return interval (ARI) of 100 years for durations greater than three hours (ibid.)
4	A strong monsoon and the strongest El Niño on record (Reisinger et al., 2014)

The 2011 flood was generally caused by an emergency release upstream from the gated Wivenhoe Dam and resulted subsequently in a successful class action by the affected landholders against the dam operators (Rodriguez & Sons Pty Ltd v. Queensland Bulk Water Supply Authority and Lynette Joy Lynch v. Queensland Bulk Water Authority trading as Seqwater, 2019). The judgement delivered on 29 November 2019 suggested, in summary, that the dam operators: (i) failed their duty of care and grossly exacerbated the scale of flood through their action; and (ii) did not follow their own operational manual including forecasts of further rainfall (ibid.).

Prior to the subject rainfall event (see Table 3.3), the operator's calculations were arguably based on the water level in the dam and not the amount of rainfall forecast (ibid.). The water level in Wivenhoe Dam referred to the dam's two compartments and two contradictory functions (van den Honert & McAneney, 2011), that is:

- To act as a buffer against drought, which meant that it was desirable to keep the dam as full as possible in case future rainfall is low
- To provide a buffer against floods, which meant that it was desirable to keep it as empty as possible to maximise retention of water.

As with the subject dam release, the protection of the structural integrity of the dam was the priority to the operator (ibid.) (The dam operations procedures have changed since 2011, see Chapter 4). This stresses the importance of asset management documentation, but *a question is also arising whether it is realistic to expect planning and documenting all proper and effective asset operation procedures in the readiness for a rainfall event of such extraordinary intensity? Should the documentation be supplemented by experience and knowledge of the operator and/or event specific engineering calculus?*

An issue of poor land-use planning (i.e., planning residential, commercial and industrial developments served by water supply assets) was also raised by the post-2011 flood inquiry with the resulting relocation of residential homes from the floodplain to higher ground outside the flood zone and raising the Defined Flood Level imposed upon all new developments (ibid.). However, the larger flood events are more than certain with the climate change accelerating so that the flood zone would also need to increase. Climate resilience through proper land-use planning is surely to be considered (e.g., Dutch *room for river* approach), but *a question arises how to balance the required flood zone extent with the population growth?* (Traditional building standards provide some guidance on climate adaptation. For example, Queensland's traditional houses [*Queenslanders*] were built on stilts to elevate and make them immune against minor flood events in addition to the increased cooling effect during hot summers, see Close, 2020).

In addition to the flood inundation-related issues, the deterioration of raw water quality at the intake was another issue (see Fig. 3.4) with a knock-on effect on WTP operations.

Contrary to the previous flood events, the magnitude of January 2011 flood event caused a shift in water quality within the dam with a long recovery time, which was also exacerbated by the presence of fine suspended sediment particles (ibid.). *As the extreme events will occur more frequently with a greater intensity, their water quality impacts must then be considered when planning catchment management practices.* Natural

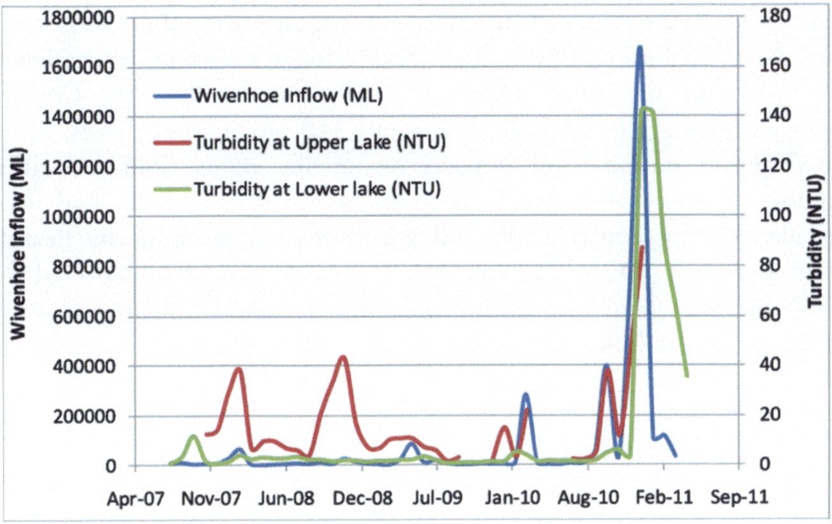

Fig. 3.4 Impacts of turbidity laden water inflow on the upper and lower sections of the Wivenhoe Dam (Watkinson et al., 2012)

catchment-related water quality issues are further discussed in Sect. 3.3 and continued in Chapter 4.

Flood in Townsville (North Queensland)

Key meteorological and hydrological conditions leading to the 2019 Townsville flood (February 2019) have been summarised in Table 3.4.

Townsville's Ross Dam peaked at 248% of its capacity and caused a release into the Ross River inundating several suburbs in Townsville

Table 3.4 Meteorological and hydrological conditions leading to the 2019 Townsville flood (JBA Group, 2019)

No	2019 Townsville flood—Key meteorological and hydrological conditions
1	Sustained and heavy monsoonal rain occurred between 27 January and 8 February 2019, which totalled 1,391 mm at Townsville Airport weather station during this period
2	Based on the total monthly rainfall for February 2019, the estimated return period of this accumulation of rainfall was estimated as a 100-year event
3	Based on gauge data from the Bohle River, the height peaked at 7 m during the February 2019 flooding that was classified as severe

(ibid.). The event caused an extensive damage to public infrastructure assets including power lines and water/wastewater transmission pipelines and networks (Deloitte, 2019). Townsville City Council reported breakage to 21 bulk mains and 310 reticulation mains, inundated pump stations and impacts on smaller dams (ibid.). A great number of water main breaks occurred as no thrust blocks had been installed during construction and with softened ground, the flexible joints gave way (Water Source, 2019). A need was also identified for a suitable asset plan to ensure protection of critical pipelines crossing creeks and rivers (ibid.).

3.3 LOS Impact

As discussed earlier in this chapter, weather anomalies and extreme weather events might have many impacts on the level of water supply services. They include impacts on water quantity and quality and hence on safety, quality, reliability and availability of supply, environmental sustainability and last, but not least, cost.

3.3.1 Water Quantity

In general, impact on the water quantity is related to: (i) adverse natural events (e.g., flood-drought cycle); (ii) supply management (e.g., lower inflows into dam storages); and (iii) demand management (e.g., water restrictions).

Asset engineers might manage supply and demand by maximising and reducing them respectively to the extent practicable. The supply maximisation should primarily be conducted in a way that does not exacerbate climate change (that is, by minimising the use of energy, particularly if the energy is non-renewable). Therefore, the protection of natural water supply sources seems to be prudent. However, the local asset engineers can only do this through proper management of dam storages and natural catchments (or groundwater resources with their recharge zones, if used for water supply, as with Perth in Western Australia and Malta).

The relevant practices and procedures have been discussed throughout this book as *lessons learned*.

3.3.2 Water Quality

In general, raw water quality at the surface water intake can be impacted by: (i) increased turbidity (e.g., muddier runoff from catchments); (ii) wild and bushfires in the catchment (e.g., ash fallout); and (iii) contaminant concentration, algal blooms and water-borne disease outbreaks. During prolonged droughts, it might even be necessary to use brackish groundwater when surface water has become unavailable. As with water quantity, proper asset planning is necessary to reduce such risks and costs.

Increased Turbidity
Both rainfall intensity and catchment's uses are variables that might cause turbid (muddy) runoff impacting on the water intake. Conventional water treatment processes are often unable to efficiently treat raw water with high turbidity. This is particularly so if a high rainfall intensity has rarely been recorded earlier and a low water quality came as a surprise to the operator. For example, Seqwater's Mount Crosby WTPs were unable to treat the Brisbane River water because of unusually high turbidity levels in 2013 because of ex-tropical cyclone Oswald (Seqwater, n.d.).

Wild and Bushfires
Wild and bushfires within the natural catchments have a direct impact on water quality that depends on the following factors (Canning et al., 2020):

- Intensity of the fire
- Post-fire precipitation
- Catchment topology
- Local ecology.

The following contaminants can be present (ibid.):

- Low intensity fires: elevated levels of Dissolved Organic Carbon (DOC) and dissolved metals

- High intensity fires: elevated levels of suspended solids/turbidity, trace metals and inorganic nutrients (phosphorus) leading to cyanobacteria outbreaks.

Post-fire water contaminants might be divided into two main categories (ibid.):

- Priority 1—commonly occurring contaminants
- Priority 2—cyanide, Polyaromatic Aromatic Hydrocarbons (PAHs) and other less common contaminants.

Without proper monitoring and pre-treatment, even Priority 1s contaminants might cause a number of impacts on WTPs, as summarised in Table 3.5.

Contaminant Concentration, Algal Blooms and Water-Borne Disease Outbreaks
As with wild and bushfires, drought-related contaminants present at surface water intakes might include greater concentrations of turbidity, DOC, nutrients (TP and TN) and increased water temperature. They cause algal blooms (cyanobacteria outbreaks) that impact on the water treatment processes (see Table 3.5). Water-borne diseases are also projected to become more common. This is likely to have impacts on the quality of treated water with the presence of DBPs in greater concentrations and deteriorated taste and odour (aesthetic properties).

Brackish and Saline Groundwater
2019–2020 drought in Australia caused that many surface water bodies in New South Wales became concentrated and rivers and creeks went completely dry what caused the use of brackish groundwater with its distinctive taste. To overcome this problem, mobile desalination plants were deployed with desalinated water: (i) available at refill stations; or (ii) fed into the towns' water supplies (Water Source, 2020). In the future, rising sea level might also impact on the surface and groundwater quality contributing to their salinity. Seawater intrusion has already been experienced in Malta, as reported by Gourbesville (2019).

Table 3.5 Impacts of contaminants on conventional water treatment processes (Canning et al., 2020)

No	Impact	Contaminant			
		Turbidity	Total Phosphorous (TP)	Total Nitrogen (TN)	Dissolved Organic Carbon (DOC)
1	Increased coagulant use	✓			✓
2	Increased sludge production	✓			✓
3	Oxidant demand	✓		✓	✓
4	Increase Disinfection By-products (DBPs)	✓		✓	✓
5	Increase cyanotoxins		✓		
6	Increased taste and odour		✓	✓	✓
7	Increased operating costs	✓	✓	✓	✓

References

Ball, J., Babister, M., Nathan, R., Weeks, W., Weinmann, E., Retallick, M., & Testoni I. (Eds.). (2019). *Australian rainfall and runoff: A guide to flood estimation.* Geoscience Australia, © Commonwealth of Australia (Geoscience Australia).

BOM. (2019, August 28). *Archive—Forty-eight-monthly rainfall deficiency for Australia* (period—48 months, until 31 October 2009). Melbourne, Australia: Bureau of Meteorology. http://www.bom.gov.au/jsp/awap/rain/archive.jsp?colour=colour&map=drought&year=2009&month=10&period=48month&area=nat. Accessed 8 January 2021.

BOM. (2020a, January 9). *Annual climate statement 2019.* Melbourne, Australia: Bureau of Meteorology. http://www.bom.gov.au/climate/current/annual/aus/2019/. Accessed 8 January 2021.

BOM. (2020b, December 10). *Archive - Six-monthly rainfall deficiency for Australia (period – 6 months, until 31 December 2019).* Melbourne, Australia: Bureau of Meteorology. http://www.bom.gov.au/jsp/awap/rain/archive.jsp?colour=colour&map=drought&year=2019&month=12&period=6month&area=nat. Accessed 8 January 2020.

BOM. (2021, January 8). *Annual climate statement 2020.* Melbourne, Australia: Bureau of Meteorology. http://www.bom.gov.au/climate/current/annual/aus/. Accessed 8 January 2021.

C3S. (2021). *Copernicus: 2020 warmest year on record for Europe; globally, 2020 ties with 2016 for warmest year recorded.* https://climate.copernicus.eu/copernicus-2020-warmest-year-record-europe-globally-2020-ties-2016-warmest-year-recorded. Accessed 8 January 2021.

Canning, A., Deere, D., & Hill, K. (2020, January). *Bushfires and the risks to drinking water quality.* Adelaide, Australia: Water Research Australia. https://www.waterra.com.au/publications/fact-sheets/. Accessed 23 January 2020.

Close, H. (2020, February 27). Housebuilding ban on floodplains isn't enough – Flood-prone communities should take back control. *The Conversation.* London, UK. theconversation.com/housebuilding-ban-on-floodplains-isnt-enough-flood-prone-communities-should-take-back-control-132468. Accessed 3 March 2020.

Deloitte. (2019, June). *The social and economic cost of the North and Far North Queensland Monsoon Trough.* Prepared for Queensland Reconstruction Authority, Deloitte. https://www2.deloitte.com/content/dam/Deloitte/au/

Documents/Economics/deloitte-au-dae-monsoon-trough-social-economic-cost-report-160719.pdf. Accessed 23 January 2020.

Gourbesville, P. (2019). Smart water solutions for water security: From concept to operational implementation. In K. Lim, A. K. Makarigakis, O. Sohn, & B. Lee (Eds. in chief), *Water security and the sustainable development goals* (pp.47–67). Global water security issues series. Paris, France: United Nations Educational, Scientific and Cultural Organization (UNESCO) International Centre for Water Security and Sustainable Management. https://unesdoc.unesco.org/ark:/48223/pf0000367904.locale=en. Accessed 18 March 2020.

Hewitson, B., Janetos, A. C., Carter, T. R., Giorgi, F., Jones, R. G., Kwon, W.-T., Mearns, L. O., Schipper, E. L. F. and van Aalst, M. (2014). Regional context. In V. R. Barros, C. B. Field, D. J. Dbokken, M. D. Mastrandrea, K. J. Mach, T. E. Bilir, M. Chatterjee, K. L. Ebi, Y. O. Estrada, R. C. Genova, B. Girma, E. S. Kissel, A. N. Levy, S. MacCracken, P. R. Mastrandrea, & L. L. White (Eds.), *Climate change 2014: Impacts, adaptation, and vulnerability. Part B: Regional aspects* (pp. 1133-1197). Contribution of Working Group II to the Fifth Assessment Report of the Intergovernmental Panel on Climate Change. Cambridge, Cambridge University Press.

IPCC. (2014). *Climate change 2014: Impacts, adaptation, and vulnerability. Part A: Global and sectoral aspects. Contribution of Working Group II to the Fifth Assessment Report of the Intergovernmental Panel on Climate Change* (C.B. Field, V. R. Barros, D. J. Dokken, K. J. Mach, M. D. Mastrandrea, T. E. Bilir, M. Chatterjee, K. L. Ebi, Y. O. Estrada, R. C. Genova, B. Girma, E. S. Kissel, A. N. Levy, S. MacCracken, P. R. Mastrandrea, & L. L. White, Eds., pp. 1–1131). Cambridge University Press.

ISO. (2018a). *ISO 31000:2018 Risk management — Guidelines.* Geneva, Switzerland: International Organization for Standardization.

ISO. (2018b). *Asset management. Achieving the UN sustainable development goals.* Geneva, Switzerland: International Organization for Standardization, Technical Committee 251 (ISO/TC 251).

JBA Group. (2019, March 7). *One month on: A retrospective look at the Townsville flooding in February 2019.* https://www.jbarisk.com/flood-services/event-response/a-retrospective-of-townsville-flooding/. Accessed 23 January 2020.

Kovats, R. S., Valentini, R., Bouwer, L. M., Georgopoulou, E., Jacob, D., Martin, E., Rounsevell, M., & Soussana J.-F. (2014). Europe. In V. R. Barros, C. B. Field, D. J. Dbokken, M. D. Mastrandrea, K. J. Mach, T. E. Bilir, M. Chatterjee, K. L. Ebi, Y. O. Estrada, R. C. Genova, B. Girma, E. S. Kissel, A. N. Levy, S. MacCracken, P. R. Mastrandrea, & L. L. White

(Eds.), *Climate change 2014: Impacts, adaptation, and vulnerability. Part B: Regional aspects* (pp. 1267–1326). Contribution of Working Group II to the Fifth Assessment Report of the Intergovernmental Panel on Climate Change. Cambridge University Press.

Nguyen, H., Mehrotra, R., & Sharma, A. (2020, January). Assessment of climate change impacts on reservoir storage reliability, resilience and vulnerability using a Multivariate Frequency Bias Correction approach. *Water Resources Research*. Accepted on 15 January 2020. https://agupubs.onlinelibrary.wiley.com/doi/abs/, https://doi.org/10.1029/2019WR026022. Accessed 12 February 2020.

NOAA. (2020b, January 3). *Catastrophic wildfires in southeastern Australia in 2019–20*. Washington, DC: National Ocean Service. www.climate.gov/news-features/event-tracker/catastrophic-wildfires-southeastern-australia-2019-20. Accessed 12 February 2020.

Rabiee, H., Khalilpour, K. R., Betts, J. M., & Tapper, N. (2019). Chapter 13, Energy-water nexus: Renewable-integrated hybridized desalination systems. In K. R. Khalilpour (Ed.), *Polygeneration with polystorage for chemical and energy hubs for energy and chemicals* (pp. 409–458). London: Academic. Available online on 30 November 2018. www.sciencedirect.com/science/article/pii/B9780128133064000136. Accessed 6 February 2020.

Reisinger, A., Kitching, R.L., Chiew, F., Hughes, L., Newton, P. C. D., Schuster, S. S., Tait, A., & Whetton, P. (2014). Australasia. In V. R. Barros, C. B. Field, D. J. Dokken, M. D. Mastrandrea, K. J. Mach, T. E. Bilir, M. Chatterjee, K. L. Ebi, Y. O. Estrada, R. C. Genova, B. Girma, E. S. Kissel, A. N. Levy, S. MacCracken, P. R. Mastrandrea, & L. L. White (Eds.), *Climate change 2014: Impacts, adaptation, and vulnerability. Part B: Regional aspects* (pp. 1371–1438). Contribution of Working Group II to the Fifth Assessment Report of the Intergovernmental Panel on Climate Change. Cambridge University Press.

Rodriguez & Sons Pty Ltd v. Queensland Bulk Water Supply Authority and Lynette Joy Lynch v. Queensland Bulk Water Authority Trading as Seqwater. (2019). 2014/200854 and 2016/373183, New South Wales Supreme Court, Court judgement delivered on 29 November 2019. www.supremecourt.justice.nsw.gov.au/Pages/sco2_classaction/floods.aspx. Accessed 17 January 2020.

Roorda, J. (2020, January 27). Bushfires – Fighting back with an Asset Management Plan. Talking Infrastructure. http://talkinginfrastructure.com/. Accessed 28 January 2020.

Sapiano, M. (2018, August 30). *Interview - Malta: Water scarcity is a fact of life*. Copenhagen, Denmark: European Environment Agency. https://www.eea.europa.eu/signals/signals-2018-content-list/articles/interview-2014-malta-water-scarcity. Accessed 3 February 2020.

Seqwater. (n.d.). *Fact sheet – Realities of Rain – What is a flood?* Ipswich, Australia: Seqwater. https://yourseqwater.com.au/37596/documents/86232. Accessed 3 February 2020.

Seqwater. (2017a, March). *Queensland bulk water supply authority, trading as Seqwater: Water for life South East Queensland's Water Security Program 2016–2046*. Version 2. Ipswich, Australia: Seqwater.

Seqwater. (2019, May 21). *Lessons from the millennium drought*. Ipswich, Australia: Seqwater. https://www.seqwater.com.au/news/lessons-millennium-drought. Accessed February 2020.

van den Honert, R. C., & McAneney, J. (2011). The 2011 Brisbane Floods: Causes Impacts and Implications. *Water, 2011*(3), 1149–1173.

Water Source. (2019, April 24). How did the 2019 Townsville floods affect the city's water infrastructure? An on the ground look at the 1 in 500 year monsoon rain event. *Water Source*. Sydney, Australia: Australian Water Association. https://watersource.awa.asn.au/business/assets-and-operations/how-did-the-2019-townsville-floods-affect-the-citys-water-infrastructure/. Accessed 29 January in 2020.

Water Source. (2020, January 28). Mobile desal plants to treat brackish water in regional NSW. *Water Source*. Sydney, Australia: Australian Water Association. https://watersource.awa.asn.au/community/engagement/mobile-desal-plants-to-treat-brackish-water-in-regional-nsw/. Accessed 29 January in 2020.

Watkinson, A., Volders, A., Smolders, K., Simms, A., Olley, J., Burford, M., Stratton, H., Gibbes, B., & Grinham, A. (2012, April). Source water protection for SEQWater. Novel techniques to assess the effectiveness of management intervention and prioritise action. Sydney, Australia: Australian Water Association. *Water,* 30(2), 100–105.

Wong-Parodi, G., Fischhoff, B., & Strauss, B. (2015). Resilience vs. adaptation: Framing and action. *Climate Risk Management,* 10, 1–7. Amsterdam, The Netherlands: Elsevier B.V.

World Economic Forum. (2020). The Global Risks Report 2020. Insight Report.15th Edition. Geneva, Switzerland: World Economic Forum in partnership with Marsh & McLennan and Zurich Insurance Group. http://www3.weforum.org/docs/WEF_Global_Risk_Report_2020.pdf. Accessed 18 March 2020.

Xu, Z., Zhang, L., Zhao, L., Li, B., Bhatia, B., Wang, C., Wilke, K. L., Song, Y., Labban, O., Lienhard, J. H., Wang, R., & Wang, E. N. (2020). Ultrahigh-efficiency desalination via a thermally-localized multi-stage solar still. In *Energy & Environmental Science*. Royal Society of Chemistry. Accepted on 15 January 2020. https://doi.org/10.1039/c9ee04122b. https://pubs.rsc.org/en/content/articlelanding/2020/EE/C9EE04122B#!divAbstract. Accessed 13 February 2020.

4

Design and Operational Considerations for Water Supply Assets

Abstract Asset engineers have a role to play in making their water supply assets adaptable and more resilient to changing climate. Relevant measures fall generally into several risk treatment categories, that is Protection-Accommodation-Retreating/Avoiding (i.e., PARA framework). Protection against climate risks involves presently, for example, reliance on large scale and then costly infrastructure to protect against floods, droughts and fires. Risk accommodation may include smaller and local (or even relocatable and adaptable) desalination plants that are normally more affordable. Extensive use of Purified Recycled Water is also part of this risk treatment strategy. The parallel retreat/avoid strategy supplements risk accommodation. It may involve, amongst other things, management of Non-Revenue Water, Combined Sewer Overflows and WWTP bypasses, which is discussed in subsequent chapters in a context of Water 4.0's technologies and solutions. At present, asset design and operational guidelines are still mostly included in various engineering standards and codes that are based on long-term and statistically derived environmental criteria. However, the use of such criteria with the current practice of safety margins will have less relevance with climate change

R. Kijak, *Water Asset Management in Times of Climate Change and Digital Transformation*, Palgrave Studies in Climate Resilient Societies,
https://doi.org/10.1007/978-3-030-79360-9_4

accelerating. When the prescriptive standards and codes become obsolete in the future, engineering practice notes and case studies will perhaps become a more important tool for asset planners, designers and operators in the not-so-distant future. A relevant engineering body of knowledge being developed should particularly focus on digital tools and models. Unlike the use of simple environmental criteria from the standards and codes, this will also require asset engineers to expanding their own knowledge and carrying out more scientifically advanced calculations. With Machine Learning (ML) and other forms of AI readily available in the not-so-distant future, it might be prudent to leave such scientifically advanced calculus to AI with asset engineers having more conceptual and strategic roles. The remaining part of this chapter deals with current design and operational considerations that remain generally focused on the protection against climate risks rather than encompassing all available options in accordance with the PARA framework. These alternative options seem to be used only marginally and hence only briefly discussed in this chapter. This discussion is followed by a more comprehensive elaboration of current practices in Australia and Malta involving three major asset classes: seawater desalination, water recycling, water transmission and distribution. Management practices for water storages and natural catchments (natural assets) supplement this discussion.

Keywords PARA framework · Protect · Accommodate · Retreat · Avoid · Asset design and operation · Machine Learning · Seawater desalination · Water recycling · Water transmission · Water distribution · Water storages · Natural catchments · Natural assets

As already discussed in Chapter 3, climate risks can be assessed and treated by consideration of vulnerability, hazards, exposure, adaptation and resilience. Three initial factors assist with risk assessment, but engineers' role in reducing climate risks is minimal or none. They have a greater role in risk treatment, that is, making their water supply assets more adaptable and resilient to changing climate. Relevant measures fall generally into the following risk treatment categories: Protection-Accommodation-Retreating/Avoiding (i.e., PARA

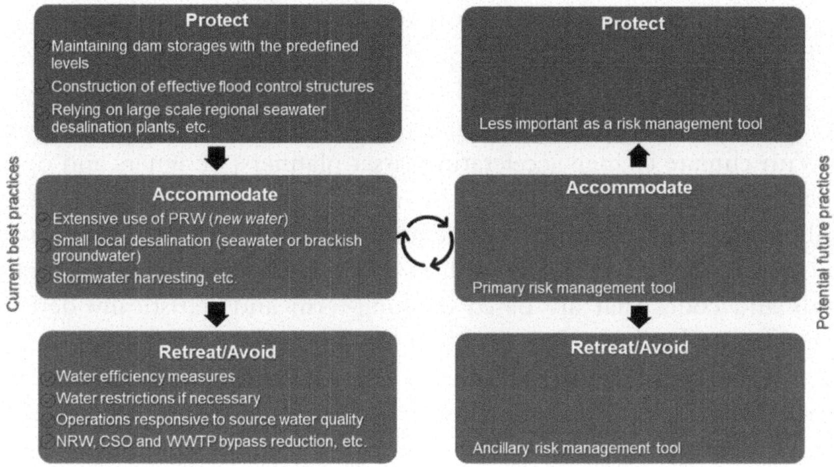

Fig. 4.1 Risk treatment categories with the current contextual examples (left hand side flowchart: current best practices; right hand side flowchart: potential future practices). The potential future practices have been illustrated in Fig. 4.2

framework—see Doberstein et al., 2019, for example). These categories with several contextual examples have been depicted in Fig. 4.1. These risk categories should be used in an iterative way following on-going risk assessment, as hazards, exposure and vulnerability change over time. Climate resilience changes accordingly.

Protection against climate risks involves presently, for example, relying on large scale and then costly infrastructure to protect against floods, droughts and fires. Risk accommodation can include smaller and local (or even relocatable and adaptable) desalination plants that are normally more affordable. This risk category involves also extensive use of PRW. It might include additional water treatment processes when raw water quality changes due to low water level (contaminant concentration) or ash fallout from bushfires at the intake. In general, the risk accommodation strategy is less costly and more responsive what is important with the increased rate of global warming. The parallel retreat/avoid strategy involves proper siting criteria for new developments to make them more immune against flood inundations or planning for off-specification turbidity levels at the intake. It includes also management of NRW, that is pipe leaks and bursts, CSOs and WWTP bypasses impacting on both quality (water intake) and quantity (recycled water volume), what is further discussed in this book in a context of Water 4.0's technologies and solutions (see Chapters 5 and 6).

With climate change accelerating, asset planners, designers and operators must learn to work with risks (uncertainties) and build up assets' resilience generally through risk accommodation. At present, asset design and operational guidelines are still mostly included in engineering standards and codes that are based on long-term and statistically derived environmental criteria (Rayner, 2019), for example, peak flood flow and other flood characteristics. However, the use of such criteria with the current practice of safety margins will have less relevance with climate change accelerating (ibid.). Safety margins would be too large with the future magnitude of climate uncertainties. Consequently, environmental criteria might not be able to capture many extreme events (ibid.). "Knowledge of the past is no longer a valid basis for making projections about the future" (ibid.).

To continue protection against risks (where feasible), the problem with asset design and operational guidelines could be partly overcome by reducing the time horizon for the relevant environmental criteria or climate variables (e.g., precipitation) used to derive these criteria. In this instance, the IPCC Fifth Assessment Report has been using a 20-year time horizon (Hewitson et al., 2014), but perhaps even a shorter time horizon might be necessary in the future. However, the life cycle and depreciation period for water supply assets is typically longer when condition-based depreciation is applied. Therefore, on-going asset planning is required with frequent asset upgrades (redesign or adaptation), as time passes. It is suggested that such process should reflect the changing climate science and up-to-date climate change projections. An alternative to the frequent asset upgrades is the process of optimising assets and minimising *waste* so that their OEE could be maximised. This might be attained through digital transformation including Artificial Intelligence (AI).

With prescriptive standards and codes becoming obsolete in the future, engineering practice notes and case studies will perhaps be a more important tool for asset planners, designers and operators in the

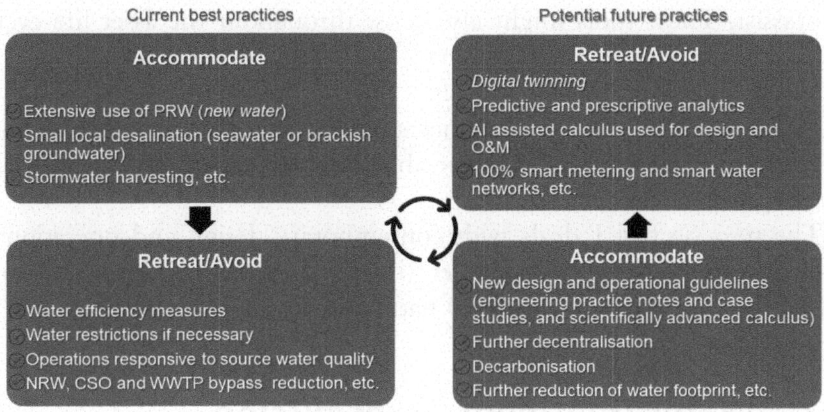

Fig. 4.2 An iterative accommodate-retreat/avoid process (Note that it is only the first iteration for the water stressed countries. The left hand side boxes contain the current best practices for water utilities, which are likely to become more common in the near future. The right hand side boxes contain the potential future practices)

not-so-distant future. *The risk accommodation strategy involves primarily asset adaptation. The risk retreat/avoid strategy is based on asset optimisation* (see Fig. 4.2). For this purpose, a body of knowledge being developed should particularly focus on digital tools and models (as with case studies discussed in this book). Unlike the use of simple environmental criteria from the standards and codes, this will also require asset engineers expanding their own knowledge and carrying out more scientifically advanced calculations (engineering calculus). The boundary between engineering and science, and science disciplines, is likely to blur. Even if assisted by advanced engineering software, this might be a difficult task, as with Bayesian stochastic methods replacing the simple *rational method* in the 2019 edition of Australian Rainfall and Runoff (ARR) guideline (Ball et al., 2019).

With Machine Learning (ML) and other forms of AI readily available in the not-so-distant future, it might be prudent to leave such scientifically advanced calculations to AI. It does not need to be somewhat futuristic AI and ML. The current tools such as Building Information Modelling (BIM) and more advanced forms of *digital twin* models might also assist. These tools might also assist throughout the asset life cycle including Operation and Maintenance (O&M), see Chapter 7.

When the engineering calculus, *digital twinning* and predictive analytics is left to AI, asset engineers will have a more conceptual and strategic role. Fewer *hurdles* in the absorption of disruptive digital tools are then expected in the future.

The next Sect. 4.1 deals with contemporary design and operational considerations for three asset classes: (i) water treatment and transmission assets; (ii) water storages; and (iii) natural assets (i.e., catchments).

4.1 Water Treatment, Transmission and Distribution Assets

4.1.1 Resilience and Adaptation for Water Assets

Water supply processes are determined by availability of raw water (quantity) and its quality, that is, by floods, droughts and fires in the climate change context. Heatwaves and increased summer temperatures have only minimal quantitative and qualitative impacts. They constitute, however, an occupational health and safety risk to the water utility staff and might have hence an indirect impact on the water supply (e.g., staff availability), but only if the respective processes have not been automated and operated remotely.

When the raw water quantity has been reduced with levels of dam storages (and other water intake sources) substantially lowered, there are several potential options for asset engineers to consider:

1. Seawater desalination, if justified by the proximity of seacoast
2. Direct or indirect water recycling with WWTP effluent being further treated to comply with drinking or other water quality standards, if acceptable to the customers
3. Switching to groundwater resources as long as they provide a sufficient quantity and quality, with: (i) fixed or mobile desalination plants (deployed permanently or temporarily) if groundwater is brackish or

saline; and/or (ii) different water treatment technologies if contaminants other than salinity are present in the groundwater
4. Purchasing and carting drinking water from a different water utility company or a local government if this option is realistic for more remote or otherwise isolated communities.

The above four options might have a high cost so a national or state (or provincial) government assistance is often necessary. Prior planning through various asset plans or water security plans before their future implementation results typically in the overall cost reduction.

4.1.2 Water Treatment, Transmission and Distribution

Non-conventional water treatment processes are further discussed in Box 4.1 in the context of Australian and Maltese experiences to ensure the required LOS and mitigate climate risks. (Non-conventional water treatment processes, as referred to in this book, include primarily seawater desalination and water recycling to produce PRW in Australia or *new water* in Malta).

In general, the focus is on the protection against climate risks rather than all available options in accordance with the PARA framework, which are considered only marginally. These options are further explored based on case studies in Chapter 6. Both Australian and Maltese water utilities could, however, be considered as model examples with respect to the NRW reduction. Their performance that is currently determined by Economic Level of Leakage (ELL)/Infrastructure Leakage Index (ILI), could further improve with disruptive Water 4.0 technologies and solutions (see Chapter 6). In fact, the English and Welsh regulating authority (Ofwat) considers no longer ELL as relevant as it does not support the industry efficiency, innovation and leakage reduction targets (Ofwat, 2020).

It is worthwhile noting that the NRW reduction is also clearly related to water efficiency. The less water is transmitted, the lower NRW is.

NRW management and water efficiency should then be considered in parallel.

The Box 4.1 contains a summary of practices in Australia and Malta involving three major asset classes: seawater desalination, water recycling, water transmission and distribution.

Box 4.1 Water Asset Practices in Australia and Malta

Seawater Desalination

Australia

- Seawater desalination (reverse osmosis, RO) is an established technology and used in all Australian capital cities on the southern Australia's coast.
- Australian desalination plants that supply water in major cities are centralised and regional scale facilities.
- Australia's first desalination plant was established in Perth (Kwinana) in 2006 followed by Gold Coast (Tugun) serving the whole Southeast Queensland in 2009, Sydney (Kurnell) in 2010, Melbourne (Wonthaggi) 2012, Adelaide (Port Stanvac) in 2012 and a second plant in Perth (Binningup) in 2012 (Wikipedia).
- Seqwater's Water Security Program proposed building two additional expandable and adaptable desalination plants (*northern and central desalination plants*) and upgrading the existing *southern desalination plant* (Gold Coast) in Queensland (Seqwater, 2017a).
- There is also a pipeline of other projects including several smaller plants throughout Australia (Wikipedia).

Malta

- Since 1982, seawater desalination has been an established technology on the Maltese Islands and currently provides around 60% of the municipal water supply (Sapiano, 2018; Water Services Corporation, n.d.)
- In Malta, groundwater resources are supplemented by water from small and local desalination (RO) plants at Pembroke, Cirkewwa and Ghar Lapsi with resulting blend of groundwater and desalinated water being stored in the 24 reservoirs (Water Services Corporation n.d.). From the security of supply viewpoint, several plants strategically located are believed to provide a greater security (ibid.)
- At present, 20% of water from the Ċirkewwa RO plant is sent via a submarine transmission pipeline to Gozo (second largest Maltese island), but there is an upgrade underway to an old RO plant on Gozo to eliminate on-going pipeline breakdowns (Water Services Corporation, 2019). Reportedly, this will also reduce energy consumption (OpEx)

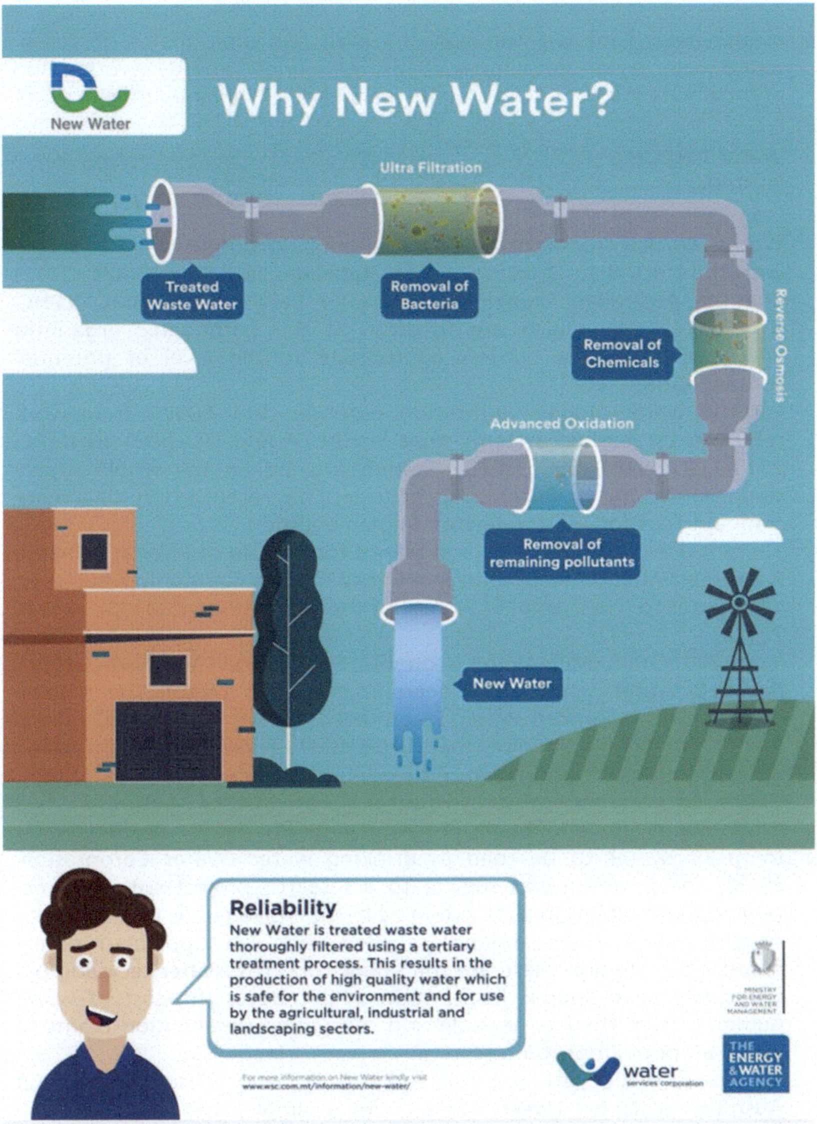

Fig. 4.3 An illustration of the polishing process for WWTP effluent to produce *new water* in Malta (Water Services Corporation n.d.)

as the new plant will use only 3.1 kWh per cubic metre of desalinated water produced (Bilocca, 2018). The planned energy consumption approximately is 50% lower as 10 years earlier (Sapiano, 2018)

Water Recycling
Australia

- Water recycling for drinking and non-drinking purposes must comply with the Australian Drinking Water Guidelines, ADWG (NHMRC, 2018) and the Australian Guidelines for Water Recycling (NRMMC, EPHC, & AHMC 2006), which are focused on the pathogenic organisms and robust multiple barriers appropriate to the level of potential contamination
- For the water quality verification purposes, the ADWG framework refers to established systems, that is HACCP and ISO 9001 (NHMRC, 2018). ISO 22000 (ISO, 2018a) system focused on the whole supply chain (*source-to-tap* including *toilet-to-tap*) has recently become a more common tool for the framework implementation
- There is a lack of full public acceptance for the use of PRW as drinking water. Seqwater (Queensland) feeds PRW to their dams only when their level drops to less than 60% as part of the drought response (Seqwater, 2017a)
- Since 2007–2008 (Millennium Drought), Queensland's Western Corridor Recycled Water Scheme has been in operation (ibid.). The scheme encompasses Advanced Water Treatment Plants at Bundamba, Gibson Island and Luggage Point that treat further WWTP effluent using microfiltration, RO and advanced oxidation processes (ibid.)
- Water Corporation (Western Australia), as part of Australia's first groundwater replenishment program in Perth, feeds recycled water to groundwater to be used as drinking water (Water Corporation, 2013). The program was subject to a research project with a three-year trial completed in 2012 followed by extensive public consultations. As with Seqwater (Queensland), recycled water can supplement desalinated water though there are still public concerns. Water recycling has reportedly an economic justification as it reduces the cost of water pumping from their large scale and centralised desalination plants to the main population centres
- For years, WWTP effluent has successfully been reused throughout Australia with no direct contact, for example, for irrigation and watering of crops, parklands or golf courses. In addition to the effluent disinfection, effluent storage security and irrigation method are key issues to minimise health and environmental risks

Malta

- Water Services Corporation's *new water* program involves an annual production of 7 million m^3 of water suitable for crop irrigation, landscaping and for industrial uses (Water Services Corporation n.d.). The project will have the capacity to potentially address up to 35% of the current total water demand of agricultural sector (ibid.)
- The aim is to achieve a net zero impact on the natural water cycle, whereby groundwater being abstracted will be replaced, directly or indirectly, amongst other things, by recycled water (ibid.)
- Currently, the *new water* program is available through hydrants, which are accessed by electronic cards with multiple hydrants available in the northern region of Malta and in Gozo (ibid.)
- The relevant polishing process is illustrated in Fig. 4.3 with WWTP effluent treated further by ultrafiltration, RO and finally by advanced oxidation
- Using shower water for toilet flushing to reduce the current average daily consumption from 110 L to approximately 70 L per person is being considered, but discussions are held on how to ensure the public health safety (Bilocca, 2018).
- The actual water efficiency trend seems to be opposite with the water consumption increasing steadily by 12% in a period of 2013–2018 (National Statistics Office, 2019)
- Recycled water has not been used so far and there are no plans for such water to be used in Malta for drinking purposes directly and indirectly

Water Transmission and Distribution
Australia

- Southeast Queensland Water Grid constitutes a network of water storages, conventional WTPs, reservoirs, pump stations, pipelines, Gold Coast Desalination Plant and Advanced Water Treatment Plants producing PRW from WWTP effluent (Western Corridor Recycled Water Scheme) with the total length of 600 km (Seqwater, 2018). In general, the grid allows for transferring water wherever it is needed in Southeast Queensland. Climate resilience is reportedly achieved through such diverse asset base and ability to use different water sources (Seqwater, 2017a)
- NRW loss in Australia is at only approximately 10% of utilities' system input for a pre-2017 period (Harris, 2018). This is reportedly due to the increased pressure through the system, poor workmanship, ageing infrastructure and soil movement (ibid.). NRW is generally driven by return on investment, in particular by ELL, which implies that if water-loss reduction activities cost more than the water is worth, the NRW reduction would be unjustified (ibid.). However more recently, reducing

NRW reduction to less than 10% is being justified based on the sustainability concerns and public health risks during drought (ibid.)

Malta

- The water supply system involves preferably smaller and local water treatment facilities to minimise the distance at which water is being transmitted.
- A smart metering system, namely Automated Meter Management (AMM), has been introduced for the whole system operated by Water Service Corporation and assists with a better interdepartmental coordination of the replacement of previous meter for smart meters, geocoding of all features of the water network, development of reports and software to analyse data received from smart meters, proper monitoring and maintenance of the whole AMM system (Gourbesville, 2019).
- The net water production was 33.5 million m^3 with leakages from water mains estimated at 3.8 million m^3 (i.e., 11.5%) in 2018 (National Statistics Office, 2019).
- Reportedly, there is a NRW control program using state-of-the-art equipment and benchmarking against best practices of other water utilities (Bilocca, 2018). This has allowed to reduce total water produced over previous years with the parallel ILI reduction. In 2015, ILI was only 1.91 compared to 1.94 attained in 2014 and 20 in 1995 (ibid.). As a result, municipal water demand has dropped by 60% since the early 1990s (Sapiano, 2018).

4.2 Water Storages

Supply and demand management measures that impact on the availability of stored water (e.g., storage level drop during drought) and storage safety (e.g., structural integrity of the over-topping dam storages) can be summarised as follows:

1. Water allocation practices within catchments of water storages
2. Replenishment of natural water resources with desalinated seawater and/or PRW

3. Water efficiency measures introduced by the consumers before financial instruments or regulatory restrictions have been applied (demand management)
4. Dam management procedures with the pre-determined water supply level and flood storage compartment
5. Evaporation control for the storages.

Regardless of whether water storages involve dams (as with southeast Queensland) or groundwater aquifers (as with Perth in Western Australia or Malta), the water overuse is the key factor that impacts on the water availability with resulting water restrictions. A responsive water allocation system might minimise the storage level drop and a water supply risk to the consumers. Desalinated and recycled waters alone are rarely able to replenish fully natural water resources due to RO's high energy consumption and relatively low public acceptance for the PRW use as drinking water. For the water allocation system to be effective, it must include enforcement of the granted water licences and unlicenced uses. To improve water use efficiency, economic instruments can also be used such as water rights transfer contracts, as with managing the Murray-Darling Basin resources in Australia (Harken & Brewster, 2019).

Water efficiency measures introduced by the consumers are to be commensurated with asset management practices. Examples included in Box 4.1 demonstrate that it has been possible to reduce voluntarily water consumption to merely 110 L per person per day in Malta or 160 L in southeast Queensland. Even a greater reduction might be possible with water grey water reuse and stormwater harvesting practices.

Unlike the above issues, dam management and evaporation control are only applicable to surface water storages so examples from southeast Queensland are relevant, where dams have presently a *full supply level* determined, that is the approved storage level for drinking and irrigation purposes (Seqwater n.d.). For their ungated dams, if inflows result in the water level rising above the full supply level, the water will spill out of the dam in an uncontrolled manner what mitigates flooding only partly (ibid.). For their gated dams, if inflows result in the water level rising above the full supply level, controlled releases will occur based on the flood preparedness and dam safety procedures (ibid.). Moreover, the ratio between the flood storage compartment and the water supply compartment must clearly be predetermined based on the hydraulic modelling results and consensus achieved. *From the liability viewpoint, it is important to document and closely follow all these conflicting objectives and corresponding operational procedures in the asset operations manuals or plans.*

There are various methods of evaporation control from water surfaces worldwide, that include (Youssef & Khodzinskaya, 2019):

- Physical methods that include continuous covers (floating sheets) or modular covers (floating objects).
- Chemical mono-layers that are single molecular layers of insoluble or sparingly soluble compounds, which when applied to water, form an invisible film.
- Floating aquatic plants are used (though water quality might become a potential concern).
- Wind breaks are made by planting trees perpendicular to the prevailing wind direction.

However, water storage coverage practices typically apply to relatively small on-farm water supply or industrial storages. Seqwater (2017a) considered them for their large dams as less favourable due to a high cost, lack of confirmed effectiveness and generally, suitability for a small area of water. Water storage coverage remains, however, a potential option that should be considered on a case-by-case basis.

4.3 Natural Assets

As most of Australia's metropolitan areas are supplied from surface water storages that receive runoff from their catchments (natural assets), the amount and quality of runoff decides about the safety and reliability of water supplies. The projected temperature increase and altered weather pattern in Australia are likely to impact on the catchments' characteristics (e.g., reduced soil moisture). As higher temperatures are drying out soil faster and a lot of water is being absorbed into soil, there is a projected reduction in the amount of runoff received by the storages (Nguyen et al., 2020). It is then important to manage the catchments to maintain the optimal soil moisture and minimise soil erosion (e.g., vegetation cover), as well as to prevent activities causing land and runoff contamination. To ensure the efficiency of investment, a whole-of-catchment approach is prudent, which considers risk-based and integrated management of both natural and built assets (Watkinson et al., 2012).

In addition to the catchment activities, natural catchment ownership is another issue to consider. Due to the ownership, natural catchments managed by water utility companies might be divided into:

- Catchments closed or open to the public
- Catchments fully or partly owned by the companies.

Unlike many other Australian catchments of water storages, most of southeast Queensland catchments are open to the public. Moreover, the local water utility company, Seqwater, is owning only 5% of natural assets (total of 1.4 million ha), which generally comprise land either underwater or directly adjacent to water (ibid.). These two issues typically elevate risk to water supplies (e.g., E. Coli contamination from cattle breeding and sediment laden runoff from land disturbance). Therefore, the relevant remediation practices should focus on revegetation, erosion control and animal control (ibid.). To remove the typical imbalance between investments in natural and built assets, Seqwater is currently investing $20 million AUD in landcare programs over the next 20 years (Water Source, 2018).

Australia's 2019–2020 *megafires* demonstrated that bushfires might be regional or even pan-regional with little regard to the local catchment practices. In this context, a question was asked about the effectiveness of previous strategies that were considered as *reactive* as no clear value had been placed on natural assets (Roorda, 2020). (See the relevant definitions of asset and natural asset in Glossary.) The *megafires* have assisted to determine such value that might be used in asset planning processes to justify catchment investments in the future (ibid.). The environmental damage caused to fauna, flora, land, water, subsoil and air can be subject to valuation. In fact, such valuation provides the cost of climate adaptation and hence permits the valuation of climate that could also be defined as an asset.

An integrated and systemic approach encompassing both built and natural assets with their clear value is prudent when the future is becoming increasingly uncertain. Based on the *lessons learned* from the 2019–2020 Australian *megafires*, the extent of water supply catchment could often be regional rather than only local (ibid.). The catchment extent has considerable implications on how the boundaries of Asset Management Systems are delineated what is further discussed in Chapter 7.

References

Ball, J., Babister, M., Nathan, R., Weeks, W., Weinmann, E., Retallick, M., & Testoni I. (Eds.). (2019). *Australian rainfall and runoff: A guide to flood estimation*. Barton, ACT: Geoscience Australia, © Commonwealth of Australia (Geoscience Australia).

Bilocca. (2018, October 21). Works to start on Gozo reverse osmosis and 'new water' network for south of Malta. *Malta Today*. https://www.maltatoday.com.mt/environment/energy/90293/works_to_start_on_gozo_reverse_osmosis_and_new_water_network_for_south_of_malta#.X6mQBVqSkph. Accessed 15 January 2020.

Doberstein, B., Fitzgibbons, J., & Mitchell, C. (2019, August). Protect, Accommodate, Retreat or Avoid (PARA): Canadian community options for flood disaster risk reduction and flood resilience. In *Natural hazards: Journal of the international society for the prevention and mitigation of natural hazards, 98*(1). Springer International Publishing AG. International Society for the Prevention and Mitigation of Natural Hazards (pp.31–50).

Gourbesville, P. (2019), Smart water solutions for water security: From concept to operational implementation. In K. Lim, A. K. Makarigakis, O. Sohn, & B. Lee (Eds. in chief), *Water security and the sustainable development goals* (pp. 47–67). Global water security issues series. United Nations Educational, Scientific and Cultural Organization (UNESCO) International Centre for

Water Security and Sustainable Management. https://unesdoc.unesco.org/ark:/48223/pf0000367904.locale=en. Accessed 18 March 2020.

Harken, B. A., & Brewster, M. M. (2019). Security and green growth: Supporting development while safeguarding water resources. In K. Lim, A. K. Makarigakis, O. Sohn, & B. Lee (Eds. in chief), *Water security and the sustainable development goals* (pp. 113–131). Global water security issues series. United Nations Educational, Scientific and Cultural Organization (UNESCO) International Centre for Water Security and Sustainable Management. https://unesdoc.unesco.org/ark:/48223/pf0000367904.locale=en. Accessed 18 March 2020.

Harris, C. (2018, December 5). Addressing non-revenue water losses. *Water source*. https://watersource.awa.asn.au/business/assets-and-operations/addressing-non-revenue-water-losses/. Accessed 6 February 2020.

Hewitson, B., Janetos, A. C., Carter, T. R., Giorgi, F., Jones, R. G., Kwon, W.-T., Mearns, L. O., Schipper, E. L. F., & van Aalst, M. (2014). Regional context. In V. R. Barros, C. B. Field, D. J. Dbokken, M. D. Mastrandrea, K. J. Mach, T. E. Bilir, M. Chatterjee, K. L. Ebi, Y. O. Estrada, R. C. Genova, B. Girma, E. S. Kissel, A. N. Levy, S. MacCracken, P. R. Mastrandrea, & L. L. White (Eds.), *Climate change 2014: Impacts, adaptation, and vulnerability. Part B: Regional Aspects* (pp. 1133–1197). Contribution of Working Group II to the Fifth Assessment Report of the Intergovernmental Panel on Climate Change. Cambridge University Press.

ISO. (2018a). *ISO 22000:2018 Food safety management systems—Requirements for any organization in the food chain*. Geneva, Switzerland: International Organization for Standardization.

National Statistics Office. (2019). *Key figures for Malta. 2019 edition. Visuals & words*. Malta, Valletta: National Statistics Office. https://nso.gov.mt/en/nso/Media/Salient-Points-of-Publications/Documents/Key%20Figures%20for%20Malta%20-%202019%20Edition/Malta%20In%20Figures%20-%202019.pdf. Accessed 18 March 2020.

Nguyen, H., Mehrotra, R., & Sharma, A. (2020, January). Assessment of climate change impacts on reservoir storage reliability, resilience and vulnerability using a Multivariate Frequency Bias Correction approach. *Water resources research*. Accepted on 15 January 2020. https://agupubs.onlinelibrary.wiley.com/doi/abs/10.1029/2019WR026022. Accessed 12 February 2020.

NHMRC. (2018). *National water quality management strategy. Australian drinking water guidelines 6 2011*. Version 3.5 updated on August 2018. Canberra, Australia: National Health and Medical Research Council.

NRMMC, EPHC, & AHMC. (2006). *National water quality management strategy. Australian guidelines 21 for water recycling. Managing health and environmental risks.* Canberra, Australia: Natural Resource Management Ministerial Council, Environment Protection and Heritage Council, Australian Health Ministers Conference.

Ofwat. (2020, May). *Reference of the PR19 final determinations: Cost efficiency – response to common issues in companies' statements of case* (p. 56). Birmingham, UK: Office of Water Services. Licensed under the Open Government Licence.

Rayner, R. (2019, January). Chapter 7 - Incorporating climate change within asset management. In C. Lloyd & M. Corcoran (Eds.), *Asset management* (2nd ed., pp. 143–161). ICE Publishing, Institution of Civil Engineers.

Roorda, J. (2020, January 27). Bushfires – Fighting back with an asset management plan. *Talking Infrastructure*. http://talkinginfrastructure.com/. Accessed 28 January 2020.

Sapiano, M. (2018, August 30). *Interview - Malta: Water scarcity is a fact of life.* Copenhagen, Denmark: European Environment Agency. https://www.eea.europa.eu/signals/signals-2018-content-list/articles/interview-2014-malta-water-scarcity. Accessed 3 February 2020.

Seqwater. (2017a, March). *Queensland bulk water supply authority, trading as Seqwater.: Water for life South East Queensland's water security program 2016–2046*. Version 2. Ipswich, Australia: Seqwater.

Seqwater. (2018). *Seqwater water grid map as at 30 June 2018.* Ipswich, Australia: Seqwater. https://www.seqwater.com.au/sites/default/files/2019-08/48125%20-%20SEQ%20Water%20Grid%20map%20-%20AS%20AT%2030%20JUNE%202018_0.pdf. Accessed 3 February 2020.

Water Corporation. (2013, August). *Groundwater replenishment scheme communications strategy 2013–2016*. https://consultation.epa.wa.gov.au/seven-day-comment-on-referrals/perth-groundwater-replenishment-scheme/supporting_documents/Appendix%2012%20Communication%20Strategy%202013%202016.pdf. Accessed 7 August 2021.

Water Services Corporation. (n.d.). *New water.* http://www.wsc.com.mt/information/new-water/. Accessed 15 January 2020.

Water Services Corporation. (2019, March 8). *Fresh water supply from Malta to Gozo restored*. http://www.wsc.com.mt/fresh-water-supply-from-malta-to-gozo-restored/. Accessed 15 January 2020.

Water Source. (2018, October 12). Seqwater pouring $1 million into protecting Brisbane River. *Water Source*, Sydney, Australia: Australian Water Association. https://watersource.awa.asn.au/environment/natural-environment/seqwater-pouring-1-million-into-protecting-brisbane-river/. Accessed 30 January 2020.

Watkinson, A., Volders, A., Smolders, K., Simms, A., Olley, J., Burford, M., Stratton, H., Gibbes, B., & Grinham, A. (2012, April). *Source water protection for SEQWater. Novel techniques to assess the effectiveness of management intervention and prioritise action. Water* 30(2), 100–105. Sydney, Australia: Australian Water Association.

Youssef, Y. W., & Khodzinskaya, A. (2019, May 29). *A review of evaporation reduction methods from water surfaces.* In E3S Web of Conferences 97, 05044 (2019), XXII International Scientific Conference *Construction the Formation of Living Environment.* https://www.e3s-conferences.org/articles/e3sconf/abs/2019/23/e3sconf_form2018_05044/e3sconf_form2018_05044.html. Accessed 29 January in 2020.

5
Defining Water 4.0

Abstract This chapter explains the relationships between *4.0's* for industry sectors and other engineering and business processes that are focused on the organisational effectiveness. For example, Water 4.0 definition used throughout this book is akin to Industry 4.0 concept applied to the water industry. Other engineering and business processes 4.0 are frequently used in parallel. In particular, Asset Management 4.0 is not about technology, but how this technology improves asset management fundamentals such as organisational culture and leadership and consequently maximises the business value. It assists with the introduction of digitalisation into the companies' ISO 55001 based Asset Management Systems. As with the fourth industrial revolution (Industry 4.0), Water 4.0 is considered in the literature as the *fourth water revolution*. Arguably, all four industrial revolutions correspond to respective four *water revolutions*. Both Water 3.0 and Water 4.0 are of importance here. They involve computers and computer-based control and the use of Cyber-Physical Systems respectively. In this chapter, relevant Water 4.0 technologies have been listed with a brief discussion followed on how Water 4.0 applies to assets and processes. In the context of Water 4.0

R. Kijak, *Water Asset Management in Times of Climate Change and Digital Transformation*, Palgrave Studies in Climate Resilient Societies,
https://doi.org/10.1007/978-3-030-79360-9_5

adaptation, the importance of strategic support from decision-makers has been identified. The decision-makers are those who are responsible for cultural changes occurring in organisations. To communicate the Water 4.0 related opportunities to them, the relevant case studies seem to be particularly helpful. Water 4.0 related opportunities and risks are finally identified in this chapter. Cybersecurity seems to be one of the greatest risks. Balancing risks and opportunities might be a challenging task in the years to come.

Keywords Industry 4.0 · Water 4.0 · Asset Management 4.0 · Maintenance 4.0 · Fourth industrial revolution · Fourth water revolution · Cyber-Physical Systems · Cybersecurity risks

5.1 Relationship Between 4.0's

Water 4.0 definition used throughout this book is akin to Industry 4.0 concept applied to the water industry (see Glossary). Other *4.0's* can be applied to other industry sectors or the corresponding engineering and business processes, as proposed by the author in Fig. 5.1.

It is worthwhile noting that Maintenance 4.0 (and other engineering and business processes) are frequently used in parallel with Water 4.0. As with Water 4.0, Maintenance 4.0's characteristics originate from Industry

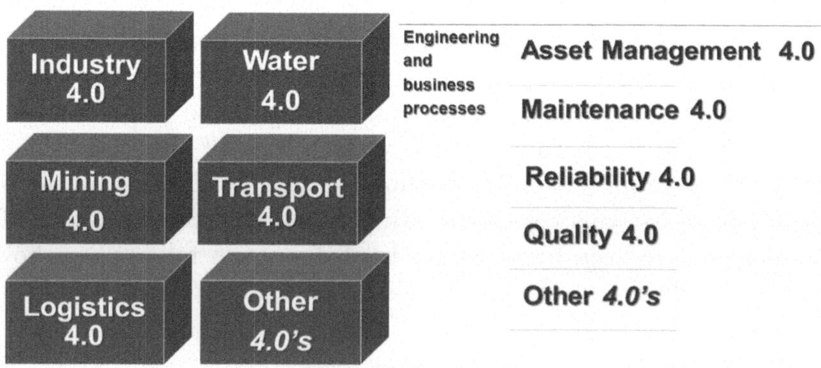

Fig. 5.1 Relationship between 4.0's for various industry sectors and the corresponding engineering and business processes

4.0 (e.g., Almagor et al., 2019). Maintenance 4.0 and Water 4.0 have thus been used interchangeably in Chapter 6's case studies involving maintenance practices for water and wastewater infrastructure.

Engineering and business processes 4.0 seem to be closely aligned with *4.0's* for industry sectors to ensure the organisational effectiveness. In particular, Asset Management 4.0 seems to be not about technology, but how the technology improves asset management fundamentals such as organisational culture and leadership and consequently maximises the business value. It assists with the introduction of digitalisation into the companies' ISO 55001 based Asset Management System (ISO, 2014b). Such approach to the engineering and business processes 4.0 is not unique and has been used earlier, for example, by Jacob (2017) in relation to Quality 4.0. All engineering and business processes 4.0 build upon traditional management methods and tools. However, they are supplemented by *connectivity*, extensive use of data (*big data*), *digital twins* and data analytics: predictive; prescriptive; and transformative (IET and Atkins). Prescriptive and transformative analytics provides recommendations on how to achieve relevant performance goals, to humans or fellow machines respectively. These issues are further discussed in Chapter 7.

5.2 Scope and Timescales of Four Water Revolutions

Is the Industry 4.0 concept applied to the water sector and Water 4.0 really the same? It seems that the term Water 4.0 was first used first by David Sedlak, a professor at the University of California and the 2014 recipient of the American National Water Research Institute Clarke Prize (Sedlak, 2015), but somewhat in a different context. He suggested that Water 4.0 should focus on the current deficiencies of centralised water/wastewater systems, wastewater reuse, seawater desalination and water efficiency measures to close the water cycle.

The initial phase of Water 4.0 as proposed by Sedlak (2015) involves upgrading the centralised systems to the current standard of those in Australia and Malta (see Chapter 4) with seawater desalination, water reuse (including *toilet-to-tap* projects) and incentivised water efficiency measures (ibid.). WWTPs will, however, evolve to fully recover water, energy and nutrients (ibid.). Water, wastewater and stormwater will be managed in an integrated way without the current artificial divisions (ibid.). The next phase will ultimately shift the responsibility for such integrated management to individual households or suburbs (ibid.). As a result, the centralised systems might ultimately be abandoned and wastewater recycled in small scale and local systems (ibid.).

As already mentioned in the previous Sect. 5.1, Sedlak (2015) definition of Water 4.0 is different than that used in this book, however. *Water 4.0 generally reflects Industry 4.0 concept applied to the water industry*, as proposed by GWP (2019). Alabi et al. (2019) suggest regardless that such concept might still assist with decentralisation.

Conflicting views are also present in literature in relation to timing of all *water revolutions*. Sedlak (2015) has proposed different definitions of the previous *water revolutions* than the more recent authors. The commencement of *first water revolution* (Water 1.0) was reported as early as 2,500 years ago with Romans building complex water pipes and channels (ibid.). Water 2.0 began in the nineteenth century with the introduction of rudimentary water treatment due to understanding of wastewater-related health risks (ibid.). Water 3.0 commenced in the twentieth century and involved construction of modern WWTPs (ibid.).

More recently, Vestner and Keilholz (2016) have proposed that *water revolutions* coincided with the corresponding *industrial revolutions* (see Fig. 5.2). Contrary to other views, Water 1.0 began only in the early eighteenth century with the use of steam and steel for high pressure structures. The subsequent Water 2.0 commenced in the late nineteenth century and was focused on electricity, pumps, turbines and hydropower rather than on water supply/wastewater treatment (ibid.).

The two initial *water revolutions* were again differently defined by the University of Exeter (as reported by Alabi et al., 2019), but without providing their timescales. Water 1.0 involved *local and* ad hoc*systems,* but Water 2.0 *large centralised infrastructure* (ibid.). Nevertheless, the definitions of more recent Water 3.0 and Water 4.0 contained in the two latter references have broadly consistent scopes and timescales. They involve computers and computer-based control (late twentieth century

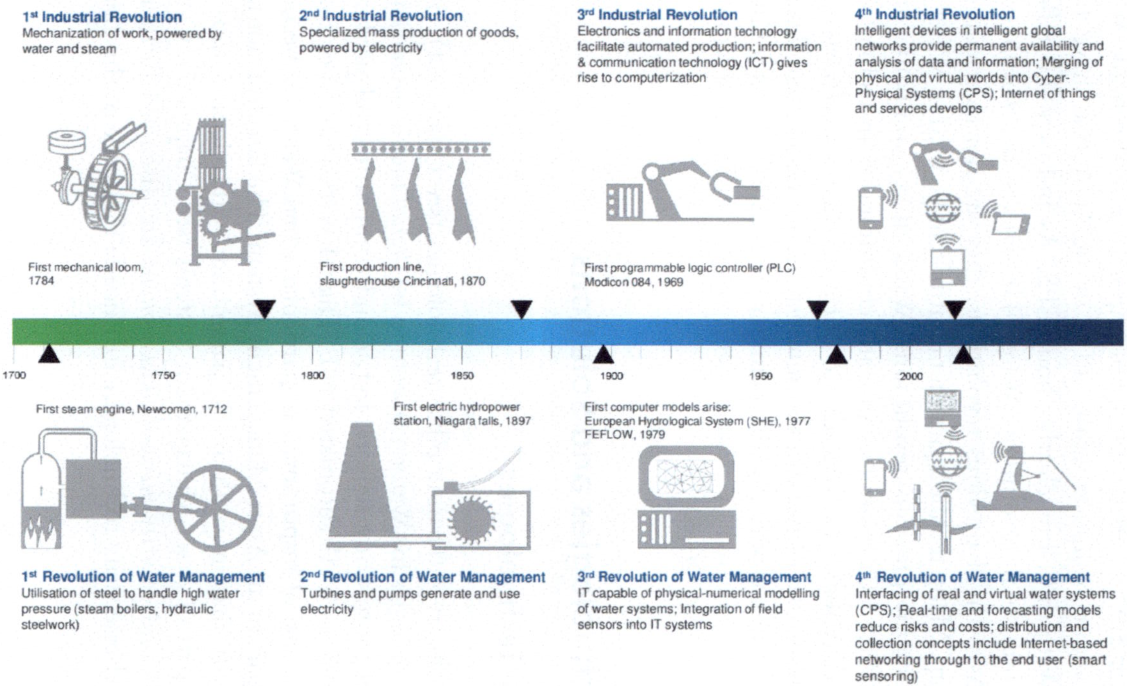

Fig. 5.2 Scope and timescales of water revolutions (Vestner & Keilholz, 2016)

onwards) and the more recent use of Cyber-Physical Systems (CPS) respectively (ibid.).

Both Water 3.0 and Water 4.0 are of importance in this book. Water 4.0 is not about computerisation, PLC or SCADA that have been subject to Water 3.0, but clearly about CPS. As Industry 4.0 reflects the fourth industrial revolution, Water 4.0 reflects the *fourth water revolution.* As with all previous revolutions, they are not clear-cut, but clearly overlap. Many technologies that continue to be used in Water 4.0 originate from the previous revolutions. Water utility companies have been using Water 3.0's remote sensing and communications technologies such as SCADA for several decades to optimise their water/wastewater treatment, transmission and distribution systems. However, digital transformation elevates reportedly the number of sensors and amount of data (*big data*) collected by these systems by 1–2 orders of magnitude (Glickman & Leroi, 2015).

5.3 Technologies and Solutions

5.3.1 Key Technologies

From the Industry 4.0's nine pillar technologies (see Chapter 1), most relevant technologies for Water 4.0 seem to be *digital twins*, visualisation, wireless asset monitoring sensors, IoT, cloud computing and predictive analytics often with AI/automated ML. Digital vertical and horizontal integration is also of importance, in particular when Water 4.0 becomes an integral part of an Asset Management System. The last, but not least is cybersecurity that must always be paramount with relevant measures included as part of Asset Management Systems (see Chapter 7). Cybersecurity risks and *security-minded management* are discussed further in Sect. 5.4. Cybersecurity technologies are outside the scope of this book.

5.3.2 Application to Assets

Asset reliability can be assessed by both ML and operator controlled systems. As with the application to processes (see below), ML can be applied in the process of anomaly detection and subsequently in a classification that identifies a cause-effect relationship and the impending asset failure (Reddy, 2020).

It makes reliability scalable and can immediately respond to multiple users with insights based on the collected data (ibid.). With ML, algorithms enable estimation of Mean Time to Failure (MTTF) and application of automated Root Cause Failure Analysis (RCFA) to eliminate the root cause (SKF, 2020). In the long term, the ML controlled system can also estimate Remaining Useful Life (RUL) (Reddy, 2020). RUL could then become the primary characteristic of predictive analytics. As reported, ML might also integrate with Computerised Maintenance Management System (CMMS) and schedule work orders and check spare parts (ibid.).

In addition to ML, there are still operator controlled hybrid systems used for data received from wireless monitoring sensors, which could also identify statistically some anomalies based on the prior asset failure history and in turn, alarm the operator. Unlike the former systems, there are unable to capture randomly occurring failures. They can only focus on the relationship between one or two pairs of features in the anomaly patterns (ibid.).

5.3.3 Application to Processes

The application of Water 4.0 to assets and processes might be somewhat intertwined, but it seems to involve process-wise the following:

- Analyses of real time and historical data through raw data acquisition, aggregation and cleaning (e.g., Newhart et al., 2019; WEF, 2017; & Xiupeng, 2013)
- The use of data and information derived from data for process controls by, amongst other things, modelling and predictive analytics, pattern

recognition and anomaly detection with the use of ML or statistics (ibid.)
- Data/information transfer through IoT based technologies (ibid.)
- Ensuring access to data/information by water utility staff and customers (e.g., WEF, 2017)
- Improved LOS to the customers (e.g., use of smart water meters) (ibid.).

For all these processes, Water 4.0 does not involve stand-alone technologies, but a combination of technologies and solutions (e.g., GWP, 2019). CPS or contextually, Cyber-Physical Water Systems (CPWS), with their *digital twins* assist with operational controls, predicting failure risks and finally, making the optimal decisions (ibid.). This is particularly so because the *digital twins*' complexity has been growing enormously (see IET and Atkins).

IoT has also an important role to play for the real time process control (e.g., GWP, 2019). With the availability of real time data, only ML is often able to make all necessary connections between multiple monitoring sensors and draw periodic conclusions, with or even without historical data (Reddy, 2020). The ML assisted analytical processes are conducted continuously with periodic conclusions that assess the risk of process malfunctioning (Alanen, 2019).

5.3.4 Importance of Strategic Support

Chapter 6 below contains a number of contextual case studies. As rightly suggested by Kolditz et al. (2019), practical case studies are of particular importance for Water 4.0 adaptation and its further advancement in the water sector. Though selection of specific solutions should ideally be made in a bottom-up process (e.g., Grievson, 2018), the overall success is primarily dependent on support from decision-makers (Kolditz et al., 2019). As always, bottom-up technical choices should be made in parallel with strategic top-down decisions (see ISO 55000, ISO, 2014a). Case studies appeal more to the decision-makers who are responsible for cultural changes occurring in organisations. Without such cultural change, digital transformation does not have a chance to succeed (*Smart Water Magazine*, 2020).

5.4 Opportunities and Risks

GWP (2019) suggests that Water 4.0's goal is operational improvements, risk reduction and sustainable management (i.e., quality, service and resource efficiency). Alabi et al. (2019) add (after other authors) that Water 4.0 would assist with improvement of ageing water infrastructure, minimisation of cost, *waste* and demand, and a greater water availability. In particular, wireless sensors connected via IoT (or SCADA) allow for real time monitoring including smart water metering, predictive maintenance, detecting and mitigating NWR (water losses), CSO and WWTP bypasses (e.g., Alabi et al., 2019). CSO and WWTP bypasses might result, for example, from pump failures and the sewerage system being overloaded. Consequently, environmental emissions particularly via CSO and WWTP bypasses could be minimised. Water quality can also be monitored in real time from *source to tap* (Sarni, 2019). This enhances perhaps the effectiveness of the current HACCP and ISO 22000 based systems when applied to water quality.

Wireless sensors installed at WTPs and WWTPs might promptly detect and respond to potential failures of assets and processes, their inefficiencies, and abnormalities (e.g., Newhart et al., 2019; Xiupeng, 2013). This reduces: (i) downtime; (ii) non-compliance with effluent standards; (iii) energy and chemicals consumption; and (iv) labour required (ibid.).

Digital transformation arguably assists with improving many centralised water systems and finding solutions to their long-term inefficiencies and failures (Alabi et al. 2019). This includes the transition from costly and inefficient reactive (or merely preventative) to predictive maintenance (SKF, 2020). The outcome thereof might be more equitable access to water and a greater system resilience (WEF, 2018). Digital technologies have also the potential to *democratise* access to water data and information (Sarni, 2019).

Digital transformation within the water sector seems to: (i) improve the customer service; (ii) optimise asset operations; and (iii) offer new products and services (Glickman & Leroi 2015). Even hydraulic models that have been used for planning purposes for decades, might be applied for operational purposes when supplemented with real time monitoring data and act as a *digital twin* (e.g., Poljak, 2018). Real time monitoring and control of water pressure, flow and water consumption with smart meters and networks have reportedly become an effective tool for reducing water losses/NRW (ibid.).

It is worthwhile noting that digital transformation is often associated with renewable energy sources (e.g., solar panels, wind turbines and anaerobic digestion), which could be monitored remotely and optimised with *big data* so that energy consumption is ultimately reduced (Microsoft, 2016). Such partial decentralisation of power supply makes the system not only more climate resilient (e.g., fewer storm-related power outages), but reduces further greenhouse gas emissions—the root cause of climate change.

Transition to Water 4.0 might also assist with decentralisation and off-grid water supply systems (ibid.), that is implementation of the original Sedlak (2015) concept of Water 4.0. Decentralisation can be understood in various ways. Grievson (2020) defines decentralisation contextually as the ability of CPS to make decisions on their own. Lall (2019) suggested that with digital transformation all local water sources might theoretically be made available, that is rainwater, surface water, groundwater and PRW. Beckett et al. (2019) highlight specifically the PRW importance in the context of Water 4.0. With Water 4.0, WWTPs might perhaps make autonomous decisions to recycle wastewater based on effluent quality data without human psychological reservations.

To sum up opportunities more strategically, they might include:

- Addressing water demand by an increased water supply (climate resilience)
- Improving environmental performance (including a lower carbon footprint)
- Potential for water supply decentralisation
- Reducing energy consumption and chemical use
- Better operational management and control (process and asset performance)

- Reducing total life cycle costs (TotEx)
- A greater water use efficiency and a lower water footprint
- Improving LOS and customer experience
- Addressing SDG 6—Clean water and sanitation and SDG 13—Climate action.

Risk-wise, it must be ensured that all collected data are of good quality to make it successful (e.g., Beckett et al., 2019; Grievson, 2018). A risk is that the collected data are wrong and consequently information extracted from such data is also wrong what might result in the future resistance to the implementation of Water 4.0 (ibid.). Moreover, it seems that the recognition of value of both data and information is still relatively low in the water industry (Grievson, 2020).

Unfortunately, there are also more obvious risks. Digitalisation transformation within the water sector operating critical infrastructure has exposed it to cyberattacks (e.g., WEF, 2018). Hacking can result, for example in:

- Stealing of customer personal information (Alabi et al., 2019)
- Disrupting the billing system (ibid.)
- Disabling the pumps, valves and other water infrastructure (ibid.)
- Changing chemical levels (Grievson, 2020)
- Human operators being misled and make erroneous decisions (Blumensaat et al., 2019).

Cybersecurity has become one of the greatest risks associated with digital transformation (e.g., Blumensaat et al., 2019; Microsoft, 2016). As a result, water utility companies often require on-premises solutions instead of cloud-based solutions (Sarni, 2019), which slows downs the progress of the transformation. Balancing risks and opportunities might be a challenging task in the years to come.

References

Alabi, M. O., Telukdarie, A., & Janse van Rensburg, N. (2019, December). *Water 4.0: An integrated business model from an industry 4.0 approach.* In The Proceedings of the 2019 IEEE International Conference on Industrial Engineering and Engineering Management (IEEM). https://www.researchgate.net/publication/339021634_Water_40_An_Integrated_Business_Model_from_an_Industry_40_Approach. Accessed 8 July 2020.

Alanen, P. (2019, February 14). How artificial intelligence is transforming the water sector: Case Ramboll. *Silo.ai.* https://silo.ai/how-artificial-intelligence-is-transforming-the-water-sector-case-ramboll/. Accessed 7 June 2020.

Almagor, D., Lavid D., Nowitz, A., & Vesely, E. (2019). *Maintenance 4.0 implementation handbook.* reliabilityweb.com. Fort Myers, FL, United States.

Beckett, R. C., Chapman, R. L., Berendsen, G., Dalrymple J., & Quispe-Chavez, N. (2019). *Quality 4.0' and water management practices.* A paper for a presentation at ANZAM 2019 conference. https://www.researchgate.net/publication/340132763_%27Quality_40%27_and_Water_Management_Practices. Accessed 7 June 2020.

Blumensaat, F., Leitao, J. P., Ort, C., Rieckermann, J., Scheidegger, A., Vanrolleghem, P. A., & Villez, K. (2019, June 7). How Urban Storm- and Wastewater management prepares for emerging opportunities and threats: Digital transformation, ubiquitous sensing, new data sources, and beyond—A horizon scan. *Environmental Science Technology, 53*, 8488–8498.

GWP. (2019). *Water 4.0. Made in Germany*. German Water Partnership.

Glickman, J., & Leroi, A. (2015). *Adapt and adopt: Digital transformation for utilities.* Bain & Company. https://www.bain.com/insights/adapt-and-adopt-digital-transformation-for-utilities/. Accessed 8 June 2020.

Grievson, O. (2018, September 31). Water 4.0 and the wastewater cycle. *WWT Magazine.* Faversham House Group Ltd 2020. https://wwtonline.co.uk/features/water-4-0-and-the-wastewater-cycle#.XJCgo7ixWM. Accessed 18 June 2020.

Grievson, O. (2020). *Is water 4.0 the future?* WEX global 2020. https://wex-global2019.com/is-water-4-0-the-future/. Accessed 18 June 2020.

ISO. (2014a). *ISO 55000:2014 Asset management—Overview, principles and terminology.* International Organization for Standardization.

ISO. (2014b). *ISO 55001:2014 Asset management: Management systems—Requirements.* International Organization for Standardization.

Jacob, D. (2017). *Quality 4.0 Impact and strategy handbook: Getting digitally connected to transform quality management*. LNS Research. https://www.sas.com/content/dam/SAS/en_us/doc/whitepaper2/quality-4-0-impact-strategy-109087.pdf. Accessed 8 June 2020.

Kolditz, O., Karsten Rink, K., Nixdorf, E., Fischer, T., Bilke, L., Naumov, D., Liao, Z., & Yue, T. (2019). Environmental information systems: Paving the path for digitally facilitated water management (Water 4.0). *Engineering, 5*, 828–832.

Lall, U. (2019). III. Positive water sector disruptions by 2030. In F. Machado & L. M. Mimmi (Eds.), *The future of water a collection of essays on "disruptive" technologies that may transform the water sector in the next 10 years* (pp. 45–61). Inter-American Development Bank, Water and Sanitation Division.

Microsoft. (2016, October). *Water industry: Benefits of moving to Cloud technology.* Prepared for Water Sector Organisations, United Kingdom. https://info.microsoft.com/rs/157-GQE-382/images/WaterSectorCloudPoV.PDF. Accessed 8 July 2020.

Newhart, K. B., Holloway, R. W., Hering, A. S., & Cath, T. Y. A. (2019, March 21). Data-driven performance analyses of wastewater treatment plants: A review. *Water Research, 157*, 498–513. Elsevier B.V.

Poljak, D. (2018). *Industry 4.0—New challenges for public water supply organizations.* https://bib.irb.hr/datoteka/940425.INDUSTRY_4.0__New_Challenges_for_Public_Water_Supply_Organizations.pdf. Accessed 8 July 2020.

Reddy, R. (2020, June 17). *Machine learning 101 in predictive maintenance.* IndustryWeek. https://www.industryweek.com/technology-and-iiot/article/21134278/machine-learning-101-in-predictive-maintenance. Accessed 8 July 2020.

Sarni, W. (2019). IV. The future of water is digital. In F. Machado & L. M. Mimmi (Eds.), *The future of water a collection of essays on "disruptive" technologies that may transform the water sector in the next 10 years* (pp. 62–75). Inter-American Development Bank, Water and Sanitation Division.

Sedlak, D. (2015). *Water 4.0: The past, present, and future of the world's most vital resource.* Yale University Press. Reprint Edition, 31 March 2015.

Seqwater. (2019, May 21). *Lessons from the millennium drought*. Seqwater. https://www.seqwater.com.au/news/lessons-millennium-drought. Accessed February 2020.

SKF. (2020). *AI for wastewater treatment Big Data's untapped potential for Water/Wastewater*. https://industrial-ai.skf.com/ai-for-wastewater-treatment/. Accessed 8 June 2020.

Smart Water Magazine. (2020). Technology against climate change: ACCIONA's solutions for more efficient management. *Smart Water Magazine*, SWM Monthly 1. https://smartwatermagazine.com/news/acciona/technology-against-climate-change-accionas-solutions-more-efficient-management. Accessed 10 July 2020.

Vestner, R., & Keilholz, P. (2016). Was bedeutet der „4.0-Ansatz " für die Wasserwirtschaft? *Gewässerschutz Wasser Abwasser, Bd, 239*, 7/1 – 7/13.

WEF. (2017). *Intelligent Water Systems: The Path to a Smart Utility*. Water Environment Federation (WEF). WSEC-2016-WP-002.

WEF. (2018, September). *Harnessing the fourth industrial revolution for water*. Fourth Industrial Revolution for the Earth Series. Geneva, Switzerland: World Economic Forum. https://www.weforum.org/reports/harnessing-the-fourth-industrial-revolution-for-water. Accessed 19 June 2020.

Xiupeng, W. (2013). *Modeling and optimization of wastewater treatment process with a data-driven approach* (PhD Doctor of Philosophy, thesis). University of Iowa, USA. https://iro.uiowa.edu/discovery/fulldisplay/alma9983777011302771/01IOWA_INST:ResearchRepository. Accessed 12 July 2020.

6

Application of Water 4.0 Technologies and Solutions

Abstract This chapter contains a number of case studies on how Water 4.0 technologies have been applied to minimise Non-Revenue Water (water losses), the likelihood of Combined Sewer Overflows and WWTP by-passes, and could minimise climate risks and ultimately enhance climate resilience. Based on the technological considerations, case studies outlined in this chapter fall into five categories. Case studies have originated from Europe, Americas, Middle East, Southeast Asia and Australia and involved a number of service providers and water utility companies. The case studies have confirmed the initial literature review findings and Water 4.0 related opportunities, inter alia, enhancement of climate resilience and consequently addressing the United Nations Sustainable Development Goals 6—*Clean water and sanitati*on and 13—*Climate action.*

Keywords Non-Revenue Water · Water loss · Combined Sewer Overflow · WWTP bypass · *Digital twins* · Hybrid analytics · Predictive analytics · Internet of things · Smart water meters · Smart water networks · Sustainable Development Goals

R. Kijak, *Water Asset Management in Times of Climate Change and Digital Transformation*, Palgrave Studies in Climate Resilient Societies, https://doi.org/10.1007/978-3-030-79360-9_6

As discussed in Chapter 5, case studies appeal to the decision-makers who are responsible for cultural changes occurring in organisations, which, in turn, are required for digital transformation to succeed (*Smart Water Magazine*, 2020). Such contextual case studies are presented in this chapter. Based on the technological considerations, case studies included in the below box (3) fall into the following five categories (Box 6.1):

1. *Digital twins*, SCADA connected sensors, hybrid analytics for process control
2. *Digital twins*, SCADA connected sensors, predictive analytics for process control
3. IoT connected sensors, cloud-based hybrid analytics for predictive maintenance
4. IoT connected sensors, cloud-based predictive analytics for sewer discharge identification
5. Smart water meters and smart water networks.

It is noted that not all details of the presented case studies are not available to the author. They rely on published information or information received directly from suppliers of products and services. Nevertheless, it seems that they confirm opportunities associated with Water 4.0 (see Chapter 5), which include sustainability and climate resilience, which has been the original proposition of this book.

Box 6.1 Water 4.0 Case Studies

Digital Twins, SCADA Connected Sensors, Hybrid Analytics for Process Control

Endress+Hauser (Germany): Technology/Solution

The main goal of Liquiline Control CDC81 (*supervisory system*) is to achieve the set levels of total nitrogen and phosphorus in parallel with minimisation of: (i) energy consumption; and (ii) chemical dosage (E+H, 2020). Process automation also rapidly reacts on: (i) measured value errors; and (ii) equipment breakdowns, e.g., blowers (ibid.).

The system uses algorithms built based on digital hydraulic models (e.g., E+H, n.d.). Based on the analysis of inflowing wastewater and several hours before its treatment in the biological reactors, the system decides proper set-points for the treatment equipment, which are optimal to reduce contaminants to the required level (ibid.). (Note that biological processes change slowly.)

As a result of continuous monitoring and signal quality check, Liquiline Control should ensure a robust WWTP work and the required effluent standards (E+H, 2020). In the case of erroneous readings and consequently a breakdown, the system sets the relevant controlling circuit in a breakdown mode and generates an error message or a warning (ibid.).

With the assistance of telecommunication or wireless technologies, Liquiline Control transfers all monitoring data to the central control room so that the process is monitored or the control algorithm parameters adjusted accordingly (ibid.). The system allows also for the use of a mobile communicator (ibid.). Figure 6.1 is a screenshot showing example nitrification process values in the automatic blower control mode.

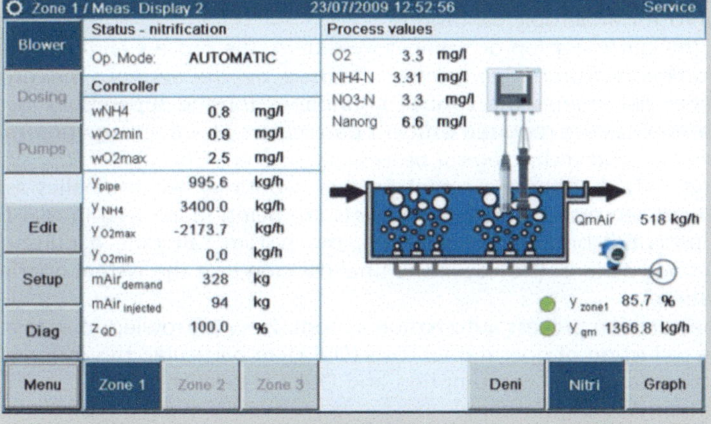

Fig. 6.1 A screenshot showing example nitrification process values in the automatic blower control mode from the Endress + Hauser's Liquiline Control CDC81 system (Świerczewska, 2020)

Endress+Hauser (Germany): Case Studies (Goal, Method, Result)

Water and Wastewater Management Company in Tomaszów Mazowiecki (Poland) (E+H, 2020*)*

Liquiline Control System ensures the full automation of WWTP processes from the inflowing wastewater pumping station and septic tank pump-out collection point. Control algorithms manage mechanical treatment processes, grit chamber, primary settling tanks, sludge thickener, secondary clarifiers and processes occurring within biological reactors.

Liquiline Control System is overseeing monitoring equipment and wastewater treatment assets. This involves an online analysis of monitoring data and estimation of set-points for assets under control. The system assesses the impact of contaminant load and ensures in a predictive way the effluent quality/parameters. Based on SCADA data, the algorithm analyses in real time the inflowing wastewater parameters and adopts the asset set-points accordingly so that conditions might be created for the optimal reduction of contaminants. Process parameters/set-points are the basis of the automatic nitrification and denitrification processes. Process parameters are determined such as aeration time and intensity, and time required for anaerobic conditions.

In the configuration of supervisory system, the process engineer's nitrification/denitrification guidelines are used by the system. The process engineer determines the control boundaries for the supervisory system. He determines the required effluent parameters at a discharge point from the reactor and technological process parameters. The supervisory system ensures only that the treatment process is optimal. All anomalies associated with sensors and treatment assets are being monitored in real time. In a potential breakdown situation, the system can take a number of predefined actions. (However, the final decision is at the system operator's discretion.)

The aeration process automation at Tomaszów Mazowiecki has resulted in the reduction of operational costs (OpEx). In particular, this includes the reduction of energy consumption and OpEx respectively.

Stadtlohn WWTP (Germany)

Stadtlohn WWTP has the capacity of ~ 21,000 EP (E+H, 2018a). The plant's two intermittent reactors are operated as sequential reactors (ibid.). Liquiline Control system controls both nitrogen reduction and phosphate precipitation (ibid.).

With respect to the automation associated with nitrogen reduction process, the operator required ensuring compliance with the regulatory seasonal standard of less than 8 or 12 mg/L for total nitrogen (TN) (ibid.). Oxygen and ion-selective ammonium/nitrate sensors were installed for each of two reactors (ibid.).

Liquiline Control System involved a load-dependent aeration control with each reactor being aerated individually (ibid.). Using the measured values, the system configures a dynamic set-point for oxygen so that an accurate amount of air required for the removal of ammoniacal nitrogen (NH_4^+-N) is supplied based on the NH_4^+-N concentration (ibid.). This should result in the process stability with the parallel reduction of blowers' energy consumption (ibid.). In addition, the system adjusts dynamically the duration of nitrification and denitrification phases to improve the reduction of TN and ensure the constant TN concentration at the WWTP discharge point (ibid.).

Liquiline Control System controls also the chemical phosphorus (phosphate) removal (E+H, 2018b. It ensures: (i) optimisation of dosing of the precipitant into the clarifiers depending on the concentration of phosphate measured at the outlet of the reactor; and (ii) a uniform phosphate concentration at the discharge point of only 0.5–0.6 mg/L for compliance with the regulatory standard of 1 mg/L (ibid.).

As required by the operator, phosphate measurements occur at three sampling points with a fully automatic phosphate analyser conducting testing of the collected samples every 10 minutes (ibid.). The system uses the measured value at the biological reactor inlet to adapt to the inflowing load (balancing out peak- and low-load phases) (ibid.).

The final compliance check with the discharge limit is at the WWTP discharge point (ibid.).

Digital Twins, SCADA Connected Sensors, Predictive Analytics for Process Control

Royal HaskoningDHV (The Netherlands): Technology/Solution

Aquasuite Virtual Operator (RHDHV, 2020a)

Virtual operator is an automated ML powered tool designed for water utilities and industries. It makes the relevant predictions (predictive control) based on historical data and external factors, and subsequently automates repetitive activities in real time (real time monitoring and control). Therefore, the virtual operator improves performance (*calm* and *smooth* network and treatment process) and allows the physical operator to focus on high-value tasks (e.g., strategic roles), which increase productivity and ultimately reduce costs (e.g., OpEx).

Virtual operator monitors, analyses, visualises (dashboards) and controls the performance of the water cycle. Based on sensor data, weather predictions and other data/information, predictions are made with 97% accuracy. *Virtual operator* consists of two components: (i) *analyst*; and (ii) *autopilot*. *Virtual analyst* provides insight from real time performance, flow, load and quality data from SCADA, PLCs, process databases and IoT sensors by comparing them with historical data. Data visualisations help the (physical) operator to focus on the right events and make informed and fast decisions. *Autopilot* is self-learning and predictive tool that can reportedly prevent a problem before it occurs by monitoring performance and reacting when necessary.

Virtual operator applies to several tools discussed below. In this book, the provided case studies focus on PURE and FLOW.

Water Treatment, Transmission and Distribution (OPIR, BURST)

Virtual operator (OPIR) predicts water demand for the relevant area and ensures the required water quantity and quality (RHDHV, 2020a). It generally lowers pressure within the network and ensures fewer pressure fluctuations what results in the reduction of energy consumption, fewer leaks and bursts (reactive maintenance events) and improved LOS (ibid.). *Virtual operator* can reduce the chemical dosage at WTP based on the water quality within the network (ibid.). *Virtual operator* (BURST) can automatically control valves/pumps when leaks occur and thus reduces water losses and other consequences (ibid.).

OPIR uses a predictive flow control as depicted in Fig. 6.2.

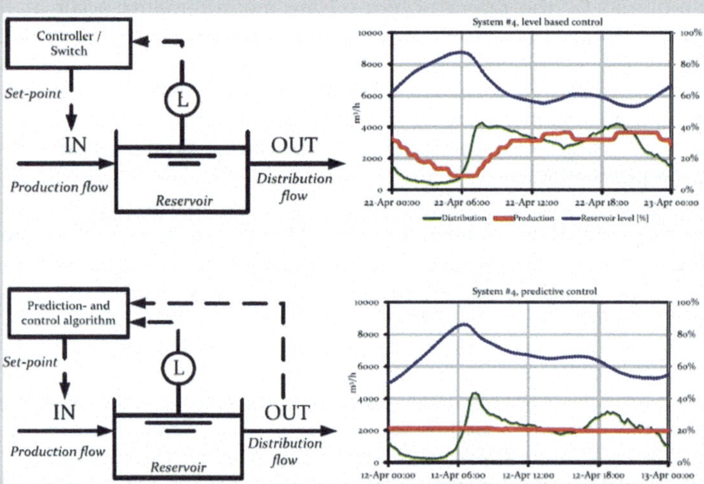

Fig. 6.2 Upper graph: Example of typical level-based flow control, Lower graph: Example of predictive flow control (Bakker et al., 2013) (Reproduced from JWSRT—AQUA, 2013, volume 62, issue 1, pages 1–13, with permission from the copyright holders, IWA Publishing)

Wastewater Collection and Transmission (FLOW) (RHDHV, 2020a)
Virtual operator (FLOW) can predict, identify and reduce CSOs. It combines weather forecasts with real time process data to predict future wastewater volumes and the implications for the network. It can identify anomalies in the sewers such as clogging and fatbergs build-up early in the process. It can monitor and control pumping stations to optimise the performance of wastewater collection systems. It can predict and reduce flow/load peaks to the WWTPs and thus the likelihood of their bypasses.
Wastewater Treatment (PURE, FLOW)
As *Virtual operator* (PURE) predicts the inflowing load to WWTP, it learns how treatment processes perform and optimises them (RHDHV, 2020a; 2020b). PURE learns the patterns of the wastewater inflow at WWTP based on data from the existing sensors and other sources (e. g. using weather forecasts, public holidays' patterns) and predicts daily inflow to anticipate the load variation (RHDHV, 2020b). As a predictive tool, it provides relevant set-points for the key wastewater treatment processes based on various criteria (e.g., energy consumption reduction) (RHDHV, 2020a). The learnt relationship between each process and the required effluent quality is used to provide predictive control set-points for each process based on the predicted influent flow and load (RHDHV, 2020b).

Though FLOW applies generally to wastewater transmission, it also assists with managing WWTP by monitoring and controlling the transmission and thus inflowing wastewater (Lubbers et al. 2018). Descriptive analytics uses data stored in a *historian* or acquired from the automation system (RHDHV 2020b). The actual performance of the sewer is assessed and compared to the reference performance from the hydraulic models (*digital twin*) (ibid.). When all characteristics are available in the asset management database, the hydraulic performance according to design can be assessed (ibid.). This gives a reference for the actual measured performance (ibid.). There is a spin-off when comparing the theoretical performance from the hydraulic model, which is based on static data (e.g., pump curves) with the actual performance from dynamic process signals (ibid.). ML is then used for fine-tuning of the control process based on *digital twins* (RHDHV, 2020b). When the analytics system has learnt the typical behaviour of the wastewater collection system, it can predict the future behaviour that is required to apply prescriptive analytics (RHDHV, 2020a).

FLOW can use a hydraulic model or a rainfall runoff prediction that allows for prediction based on the current conditions (Lubbers et al. 2018). In parallel, it uses Dry Weather Flow (DWF) prediction based on heuristic and adaptive predictive model used earlier in the water treatment optimisation (OPIR) (Lubbers et al., 2018), see Fig. 6.2.
Sludge Processing (MINE) (RHDHV, 2020a)

Virtual operator automates operational performance of sludge logistics and treatment processes. It optimises sludge treatment to maximise the production of biogas and reduce the quantity of sludge for disposal. In addition, it reportedly improves sludge handling properties.

Royal HaskoningDHV (the Netherlands): Case Studies (Goal, Method, Result)

Public Utilities Board, PUB (Singapore) (RHDHV, 2020a)

In a two-year pilot project, PURE was applied at PUB's Integrated Validation Plant (IVP). The goal was to validate: the reduction of energy consumption and chemical use; improvement in water quality; and the manpower use reduction for the water reclamation processes.

PURE was connected to the plant's SCADA system to gather data and control key processes and subsequently to send data to the Aquasuite cloud. With the data collected in *near real time* in the cloud, the software was able to track the actual performance through a *digital twin.*

More advanced analytics (e.g., predictions, anomalies, drift detection, performance indicators and virtual sensors) were made available to the operator through the cloud. This also included early warnings for specific events. In parallel, the *land-based part* of the solution was optimising the real time performance.

As reported, the preliminary results demonstrated that Aquasuite (PURE) could accurately predict and prepare for the inflowing ammonium loads several days in advance. *Autopilot* could ensure the required performance when the process was unattended. The maximum aeration flow reduction of 15% and the corresponding energy consumption savings were also demonstrated. This was related to more stable operations and improved effluent quality.

Water Authority Valley en Veluwe (WSVV) (the Netherlands) (Lubbers et al., 2018*)*

In this case study, the focus was on both predictive and prescriptive methods to improve and optimise the wastewater treatment process with FLOW. Contextually, this was associated with unsteady behaviour of rainfall peak flows entering the WWTP and low effluent quality/WWTP bypasses (or CSO) impacting on the receiving waters by causing eutrophication. To improve the effluent quality, WSVV applied earlier an effluent polishing step (i.e., tertiary treatment with sand or disc filters). However, the step's hydraulic capacity was smaller than the maximum WWTP hydraulic capacity what frequently caused its bypasses during rainfall events. Such peak flows entering the WWTP were also related to peak loads of Total Oxygen Demand (TOD) and Total Suspended Solids (TSS).

The optimisation of flow to WWTP was conducted by applying a predictive control instead of conventional level-based control at the pumping stations. Implementation of the predictive control using DWF prediction and rainfall forecast flatten peaks flows/loads. The predictive control is awaiting the future change of circumstances by forecasting the available storage capacity within the sewers to flatten the flow to the WWTP without causing additional CSO. In this instance, FLOW predictive controller was implemented as a pilot project.

DWF predictions and rainfall forecasts were the key issues for the FLOW controller. Based on these two issues along with the real time measurements of level and the pump discharge to sewer, a prediction and optimisation of the available storage was conducted, which limited the discharge. The rainfall runoff prediction is derived from a rainfall runoff model with the model input being based on the area details and the precipitation prediction.

At the beginning of forecasted rainfall event, there was no discharge limitation. After the rainfall event, when levels were below the defined critical level, the sewers were emptied as a result of a limited discharge. The incoming flow to the WWTP was optimised within a range between the maximum pump capacity and the capacity of effluent polishing step, but it was also dependent on: (i) time available before the next rainfall event; and (ii) time required to empty the sewers. In this case study, such control system was only operational immediately following a rain event. When the rain started, the adverse effect was reduced by a gradual increase of the pump's flow rate so that the risk of CSO could be reduced. After the rain, the focus was, however, on the energy consumption and treatment performance.

DWF, rainfall predictions along with the level and discharge monitoring data were an input to the FLOW controller. The used sewer storage capacity (i.e., current state) was continuously estimated based on the actual level and application of the relevant storage curves from the sewer models. The future sewer storage capacity (i.e., future state) was predicted by a volume optimisation technique.

The storage within sewers was modelled as separate reservoirs. The reservoir inflow prediction was estimated as a sum of DWF and rainfall runoff prediction. The reservoir outflow prediction was determined by a volume optimisation. For this purpose, the cumulative storage was optimised within two constraints, that is: (i) critical levels in the sewers; and (ii) need to empty the storage in the readiness for the next significant rainfall event.

In this case study, it was demonstrated that rainwater from most of the rainfall events could be discharged with flattened peaks to several WWTPs operated by WSVV, while reducing bypasses of the tertiary treatment without additional CSOs. Also, the performance of primary and secondary treatment was improved in parallel. This resulted in a lower concentration of nutrients (phosphorus) being present in the effluent along with the concurrent reduction of energy consumption and chemical dosage, and consequently OpEx.

Xylem (United States): Technology/Solution
BLU-X with the use of advanced data analytics enables the water utilities/operators to make the optimal decisions and ensure, amongst other things, climate resilience and financial affordability (Xylem, n.d.). With the understanding of external factors (e.g., weather) influencing operating parameters and the ability of analysing a great number of potential scenarios in real time, water utilities can manage uncertainties associated with asset planning processes in parallel with ensuring a cost-effective operation (ibid). Xylem refers to their approach as *decision intelligence* that is based on *big data* (ibid). The goal is to better inform the present system-level choices and make recommendations to improve future O&M and asset planning (ibid.).

BLU-X seems to have three key applications/domains (ibid):

Drinking Water Quality

- A *source-to-tap* and real time *digital twin* based on online monitoring of source waters, WTPs, and hydraulic and water quality models
- Decision support to trade-off between supply reliability, quality and energy consumption and daily operational requirements
- Simulation tools for the operators to test various operational scenarios
- Water supply system and distribution network control and optimisation.

Urban Watershed

- A real time *digital twin* of the wastewater collection system, which optimises, controls and coordinates all system assets. It can identify defects and potential failures for critical assets by leveraging condition assessment data and ML tools.
- Intelligent pumping systems provide information on real time conditions and thus can assist with maintaining their reliability.
- *Digital twins* for WWTPs can reportedly assist with process optimisation, energy consumption reduction, and increase of the process stability, throughput rate and effluent water quality.
- When integrated with real time control of wastewater collection systems/sewers, digital twins can stabilise incoming flows into WWTP.

Decision Intelligence

- Data integration, analytics and visualisation capabilities to assist the operator (water utilities) to:
- gain control of the installed intelligent systems

- (use of the collected *big data*).

Xylem (United States): Case Studies (Goal, Method, Result)
Municipal WWTP in Cuxhaven (Germany) (Xylem, 2019*)*
A large municipal WWTP in Cuxhaven, operated by EWE WASSER GmbH (EWE), intended to optimise their energy consumption.

Using Xylem's BLU-X Treatment solution, an optimisation strategy was implemented based on artificial neural network powered models of the carbon, nitrogen and phosphorus removal processes based on data from the plant's existing SCADA system. They predicted the optimal set-points to operate the aerators of five parallel biological reactors.

BLU-X provided a real time *digital twin* of the entire plant so that each process receives optimal aeration and chemical dosage inputs based on Biological Oxygen Demand/Chemical Oxygen Demand (BOD/COD).

Since EWE had no online sensors available to take real time measurements of inflowing raw wastewater concentrations, several *virtual sensors* were developed to estimate the inflowing carbon, nitrogen and phosphorus loads. Such *virtual sensors* helped EWE to operate the aeration process efficiently and for compliance with the regulatory requirements.

As a result, a 26% reduction in aeration energy usage occurred, which totals at ~ 1.1 million kWh annually, without compromising the regulatory effluent quality standards.
Evansville Water and Sewer Utility (EWSU), Evansville, Indiana (United States) (Xylem, 2020a*)*
Evansville is located on the Ohio River with the city's combined sewers comprising approximately 40% of the total sewered area. The combined sewers had a prior history of overflowing 1.8 billion gallons (6.8 million m^3) of untreated sewage annually from 22 CSO events into the receiving tributaries of the Ohio River and Pidgeon Creek. As a result, the city entered a consent decree with the Environmental Protection Agency (EPA) in 2011 to increase the sewer capacity and minimise/prevent the CSO with the City's long-term control plan (LTCP) budget of $1 billion.

After the decree was issued, the local Evansville Water and Sewer Utility (EWSU) partnered with Xylem to apply its *Decision Intelligence* approach using BLU-X Wastewater Network Optimization. The approach created a real time decision support system (RT-DSS) based on a data-driven Xylem's *Sense-Predict-Act* methodology.

EWSU intended to initially understand how the sewerage system behaved based on real time data and the system model. For this purpose, the *Sense-Predict-Act* methodology was applied, that is:

- *Sense* was used for real time monitoring of their sensor networks to gather and integrate all critical level, flow, and rain gauge data.
- *Predict* involved data integration with the previous hydraulic models by application of Machine Learning to develop a digital twin of EWSU's sewer network. dynamics for their entire collection system.
- *Act* was used for forecasting future outcomes based on the analysis of real time data with the real time operational model. These predictions allowed EWSU staff to make decisions on optimisation of the existing sewer assets.

Opportunities were identified on how to cost-effectively improve system performance. By using a real time decision support system approach to manage their existing sewer assets, it was possible to fine-tune operational responses during wet weather events (see Fig. 6.3). This resulted in CSO being reduced by more than 100 million gallons (380 thousand m^3) per annum by using the real time decision support system approach to manage the existing sewer assets and optimise the CSO gates and lift station.

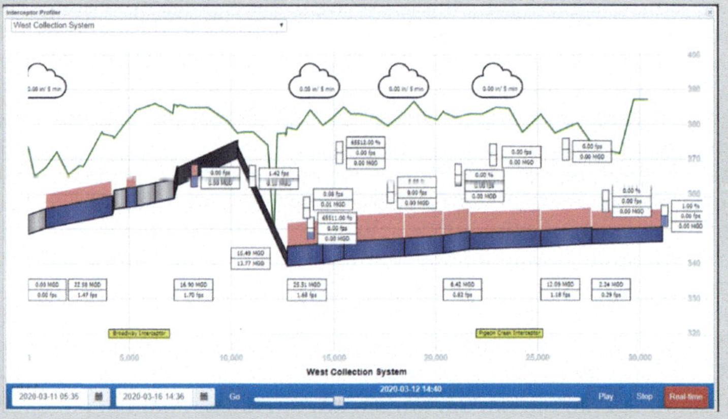

Fig. 6.3 BLU-X Wastewater Network Optimization monitoring hydraulic grade lines, CSO outfall and control gates at Evansville

In addition, the RT-DSS was able to show well levels and available capacity at the WWTP (within the next 30–60 minutes) during significant wet weather events and provided pumping rate recommendations to eliminate *guesswork*.

ReliaSol (Poland): Case Study 1's Technology/Solution (Lipnicki, 2021)

- Neural Network (NN)-based models have been applied in the below case studies. The main advantage of NN-based models is the ability to handle large non-linear, non-stationary data sets without the need to define dependencies between the input variables. Artificial NN can identify these dependencies and determine the result. Arguably, another advantage is the fact that a person designing the network for a problem being resolved does not need to have an extensive knowledge in the problem area.
- Data included historical data from SCADA and Distributed Control System (DCS) and real time data. The locally integrated data were subsequently sent to the cloud.
- For the assessment of model validity, data is divided into training and testing sets. The training set is used to build the model, while the testing set, is used only to assess its predictive capacity. The testing set is not used for training. This approach permits to reliably assess the effectiveness of the network in the real conditions.

ReliaSol (Poland): Case Study 1's Goal, Method, Result (Lipnicki, 2021)

The goal was to develop a model for an undisclosed water utility in Poland, which would predict water demand/consumption every 10 minutes. Ultimately, the model provided predictions for a 24-hour time horizon for each 10-minutes interval. The water consumption was understood by the customer as the sum of changes in the flow meter readings for nine pumps.

The data set provided by the customer included readings every 10 minutes from sensors installed at the pumps for a period of six years. Data included pressure and flow measurements. They were supplemented by weather data and information on public holidays and other events.

The water demand predictions were intended to focus on the following two scenarios: (i) the demand exceeds substantially the supply; and (ii) the supply exceeds substantially the demand. Both scenarios could result in unplanned financial costs for the utility. In scenario (i), the costs would be a result of penalties for failing to deliver the contracted water volume. In scenario (ii), such costs would be caused by the oversupplied and subsequently wasted water.

To forecast the demand, it was decided to apply classical regression statistical models, two types of neural networks and gradient enhancement of decision trees (gradient boosting). In addition to raw data, various time-based statistics were estimated and fed into the model (e.g., 24-hour average and maximum last hour value).

In this case, data from the initial five years were used as the training set, and the data from the last year as the testing set. This approach assumed that if the model generalised knowledge well for the most recent data, it will also perform well in real conditions.

Another issue was choosing an appropriate metric to evaluate the quality of the model. It often happens that one metric is not sufficient to reliably determine the effectiveness of the model. In the case of predicting water demand, the following two metrics were used to evaluate the models:

1. Number of predictions for which the relative error was below 5% (main metric)
2. Coefficient of determination, R^2.

When the model had been built, the results for the testing set were as follows:

1. Approximately 97% of water demand predictions had an error of less than 5%
2. The coefficient of determination was above 0.99.

With approximately 97% of predictions having accuracy of greater than 95%, less accurate predictions were attributed to soccer matches, social and cultural events and water losses (NRW). In general, the model allows for adequate water supply planning and optimisation.

ReliaSol (Poland): Case Study 2's Technology/Solution (Lipnicki, 2021)

- The optimisation system involves the use of a data-driven *digital twin* (linear regression) for a water treatment process.
- Before building algorithms for the optimisation system, the good practice recommendations are applied and the quality of data assessed for this purpose.
- The optimisation is only possible when the state of repair for the subject assets is well known. The optimisation process should then be aligned with the respective current maintenance strategy.

ReliaSol (Poland): Case Study 2's Goal, Method, Result (Lipnicki, 2021)

The case study involved data analysis for the optimisation of coagulant consumption in the drinking water treatment process. The main goal was to reduce the consumption of coagulants and associated OpEx, while maintaining water quality indicators related to water turbidity. The dosage of the coagulants is determined by its type, raw water quality and induced settling properties.

Prior analyses for dosing of coagulants were required to ensure that the process could be optimised. The analyses included the following:

- A preliminary process data analysis including information on data gaps and initial statistical analysis
- A preliminary check of the adopted optimisation algorithms.

In general, the available data comprised parameters of coagulants, water quality parameters (before and after coagulation) and readings from sensors for the flow through the pipeline.

Based on the available data, the optimisation system identified values of relevant control parameters, which were optimal at a point of time. The identification of these values was dependent on the sampling locations. Based on the conducted analyses, raw water and the drainage tank sampling seemed to be particularly important. Samples at these locations were then taken twice daily so that the optimisation system was able to respond twice daily. To improve the effectiveness of optimisation algorithms, it was, however, recommended to increase the frequency of sampling followed by laboratory analyses of the collected samples.

A linear relationship of the turbidity reduction was generally observed. However, the presence of discernible outliers might suggest that there was feasible to optimise the process by about 10%. In addition, the following was discovered:

- On same occasions, the final value of turbidity was significantly higher than the respective standard.
- For some turbidity samples before coagulation, differences in the respective turbidity values after coagulation were even up to 100%.
- There was a non-linear correlation with the difference in turbidity before and after coagulation, which was not always directly proportional to the dose of coagulant.

The available data was assessed in this instance as an average quality, but sufficient for the application of optimisation algorithms.

IoT Connected Sensors, Cloud-Based Hybrid Analytics for Predictive Maintenance

Dynamox (Brazil): Technology/Solution (Passos, 2020**)**

As with many other countries, considerable water losses (NRW) during transmission and distribution in Brazil adversely impacts on the energy, carbon and water footprints. The water losses result mostly from: (i) insufficient maintenance practices for the ageing water infrastructure; and (ii) the infrastructure often at distant locations what further hinders the execution of planned maintenance. To improve the maintenance practices and reduce respective impacts, Dynamox applies Water 4.0 and Maintenance 4.0's predictive technologies to water pump systems.

The below case studies demonstrate that despite locations at distances greater than 150 km, potential pump and motor failures can be identified in advance with DynaPredict solution developed by Dynamox without the presence of a technician onsite.

Centrifugal Pump Characteristics and Failure Modes

Pumps are usually classified according to the trajectory of liquid flow in the impeller: radial flow, mixed flow and axial flow. Centrifugal pumps are classified as radial flow pumps. They are characterised by a rotating component (i.e., impeller) equipped with vanes, which exerts forces on the liquid resulting from acceleration generated. Centrifugal pumps are commonly used at water pumping stations.

Water penetrates the centrifugal pump through an inlet situated near the shaft, from which is subsequently directed to the periphery at high speed due to the centrifugal force generated by the impeller. The water exits the impeller tangentially and is channelled into a tapered circular chamber, called a volute, where a part of its kinetic energy is converted into pressure.

For condition monitoring, the most important pump characteristics are the number of vanes and the type of volute. For a single stage and single volute centrifugal pump, typical failure modes such as misalignment, bearing anomalies and cavitation, have been included in tables and databases. They depend on the monitored frequencies (i.e., revolutions per minute, RPM). For a multi-stage pump, the vane pass of each stage is to be monitored. This is because each stage has a different number of vanes.

The pressure in the volute varies when the impeller rotates, and depends on how the vanes align with the outlet. Consequently, the pressure in the volute pulses at a frequency equal to the number of vanes multiplied by the shaft speed. This value is known as the *vane pass frequency*. For a double volute pump, the number of vanes used to calculate the vane pass frequency is doubled.

DynaPredict Solution
DynaPredict is a solution comprising vibration and temperature analyses for rotating equipment. It involves a wireless technology used for predictive maintenance and includes main components that are depicted in Fig. 6.4.

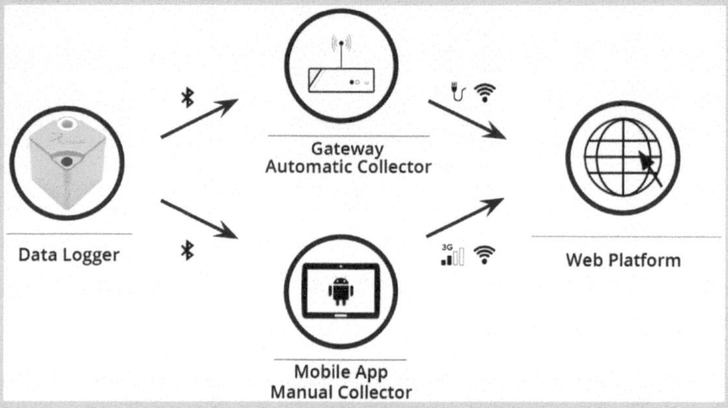

Fig. 6.4 Relationship between the main DynaPredict solution components (*Source* http://www.dynamox.net)

DynaLogger—a datalogger with a triaxial accelerometer, temperature sensor, internal memory, autonomous three-year battery life and Bluetooth data transmission. It continuously monitors equipment and provides global and spectral graphics to analyse potential failures pre-empting the functional failures.

Web Platform—cloud-based predictive analysis software allows access to the data anywhere at anytime. The web platform includes several analysis tools for understanding the vibrational signals of the equipment.

The key feature of the web platform is a Decision-Making Dashboard (DMA), which classifies the data collected from each measurement point according to pre-configured alerts. When the alerts are triggered, the status changes from green (no alert) to yellow (A1 alert) or red (A2 alert), see Fig. 6.5. The DMA objective is to support decision-making and maintenance interventions.

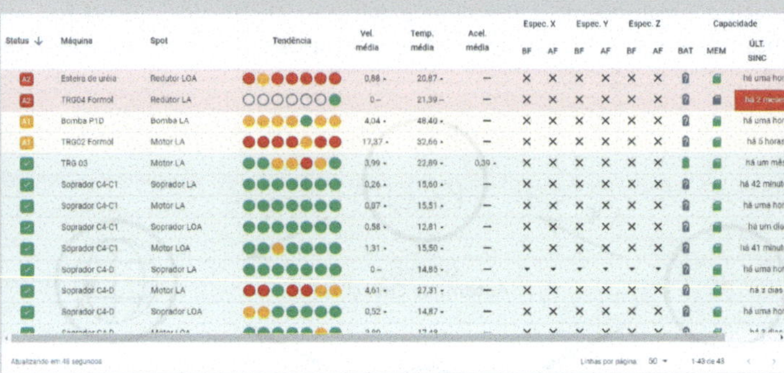

Fig. 6.5 A screenshot from of the DMA Dashboard (*Source* DynaPredict Web Platform)

It is also possible to evaluate spectra collected at each monitoring point with a chronological timeline showing when each piece of data was collected (see Fig. 6.6). This allows the user to go back in time and select a specific dataset for further analysis.

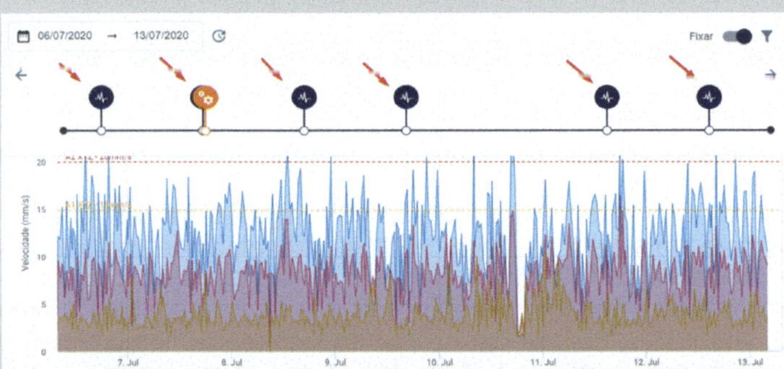

Fig. 6.6 Spectra collected over time, as indicated on the chronological timeline by the red arrows, and continuous velocity data (a graph below the timeline)

Dynamox (Brazil): Case Studies (Goal, Method, Result) (Passos, 2020**)**
The below four case studies demonstrate how the potential centrifugal pump failures have been identified and characterised (failure mode identification) with the assistance of DynaPredict solution. As a result of the undertaken maintenance practices, the required LOS to the consumers was not affected in these instances.
Case Study 1
Significant levels of vibrations at 1X RPM in the horizontal and axial directions were detected (see Adams, 2009). As a result, the system was *pre-aligned* at the base. Laser alignment equipment revealed a 6 mm horizontal misalignment and 2.5 mm vertical misalignment. Figure 6.7 depicts the vibration spectrum indicating the failure.

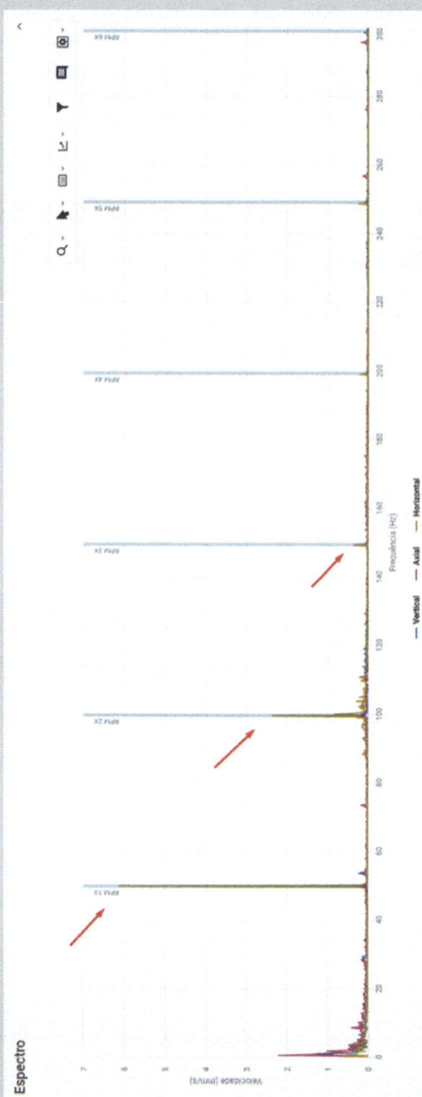

Fig. 6.7 Velocity-frequency spectrum showing peaks aligning with indicators at 1X RPM, 2X RPM, 3X RPM etc. (1 RPM = 1/60 Hz)

In the misaligned condition (see Fig. 6.8), the increase in temperature, noise and vibration dissipate some of the energy that should be converted into work, leading to a direct reduction in the efficiency of the machine (see Piotrowski, 2007).

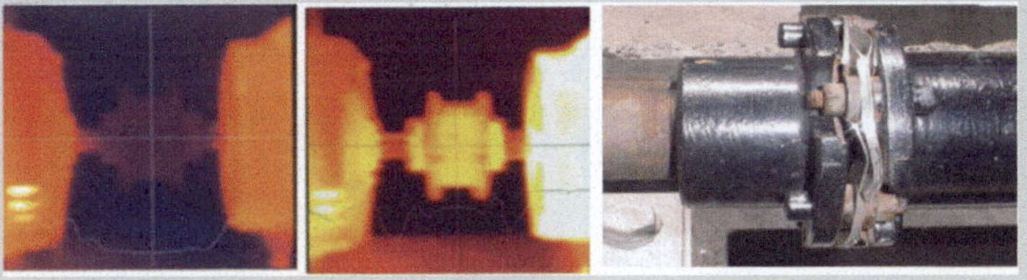

Fig. 6.8 Thermal imaging showing the difference in heat dissipation in the misaligned condition

Case Study 2
A pump system was being gradually impacted by an increase in the vane pass frequency. It was initially suspected that a problem with the vanes or casing had occurred (see Adams, 2009). However, after the pump system had been cleaned (see Fig. 6.9), the system started functioning normally and the vane pass frequencies have not been detected since. Figure 6.10 shows the relevant spectrum used for analysis in this instance.

Fig. 6.9 Image showing the component wear after the vibration identification

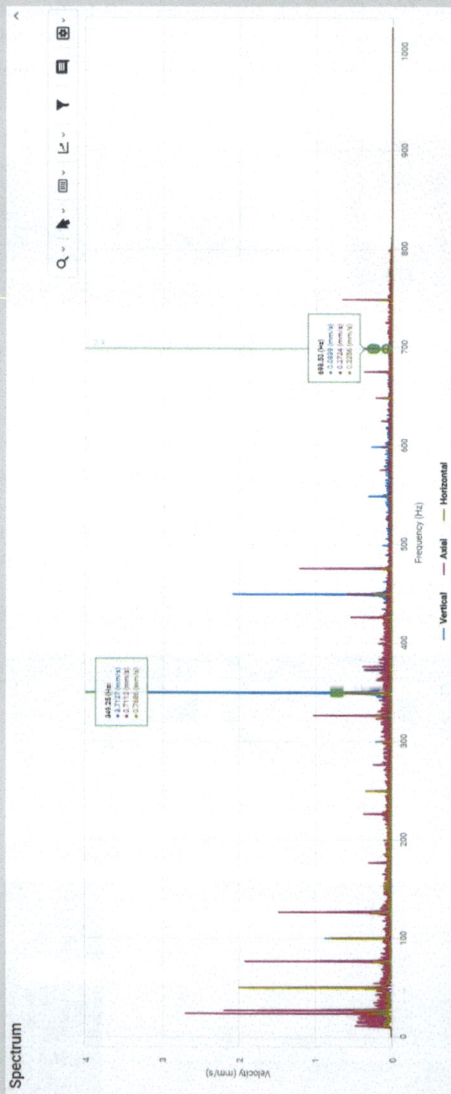

Fig. 6.10 Velocity-frequency spectrum with harmonic indicators at 350 Hz and 700 Hz

Case Study 3
In this case study, a centrifugal pump was causing high vibration levels at the operational speed. The vibration levels of the inner bearings were higher than the outer bearings, and the vertical readings were considerably higher than the horizontal readings. The structure was checked and found to be resonant in the first mode (see Mechefske, 2005). Figure 6.11 depicts the respective spectral analysis.
Case Study 4
The global measurement graphs for a centrifugal pump were indicating a simultaneous increase in vibration, expressed as *root mean square* (RMS) velocity, and temperature (see Fig. 6.12).

When the spectrum recorded at the point of simultaneous increase of both vibration level and temperature was analysed, significant peaks in the axial direction were found (see Mitchell 1978). When the equipment was correctly configured and the relevant bearing model selected from the DynaPredict database, the analysis tools showed that there had been a failure in the outer race frequency (Ball Pass Frequency Outer, BPFO) due to the lubricant's contamination with water. Figure 6.13 shows the spectrum used for analysis.

In this case study, the equipment continued to be monitored until the next scheduled downtime, when the bearing was inspected and replaced. Figure 6.14 depicts the subject damage to the bearing.

IoT Connected Sensors, Cloud-Based Predictive Analytics for Sewer Discharge Identification
Kando (Israel): Technology/Solution
As reported, Kando's Clear Upstream solution utilising *big data* and ML algorithms can provide wastewater collection network operators with real time insights into the network conditions (Steutel-Maron, 2019a). It involves monitoring of wastewater quality at sewer locations upstream from WWTPs to trace in real time anomalies and contaminants to their source (Steward, 2020). For this purpose, Kando has reportedly utilised Paperspace's Gradient ML platform (Kando, 2020a). Kando's *big data* analysis is focused on wastewater source control using live network information (Steutel-Maron, 2019a).

Fig. 6.11 Velocity-frequency spectrum showing a large spike at 50 Hz

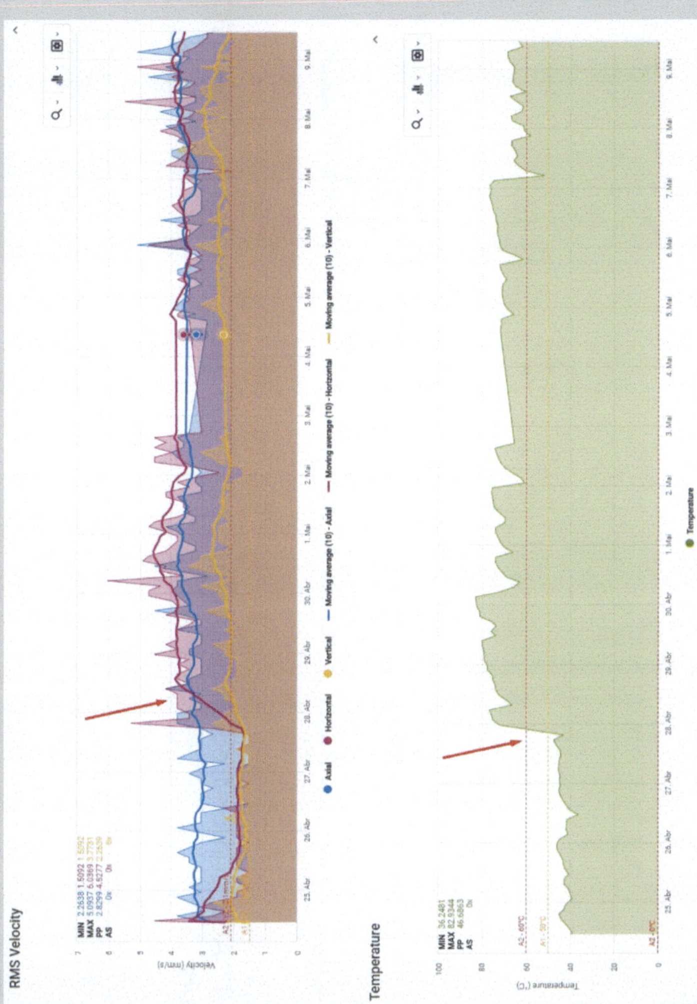

Fig. 6.12 Simultaneous increase in vibration levels (upper graph) and temperature (lower graph)

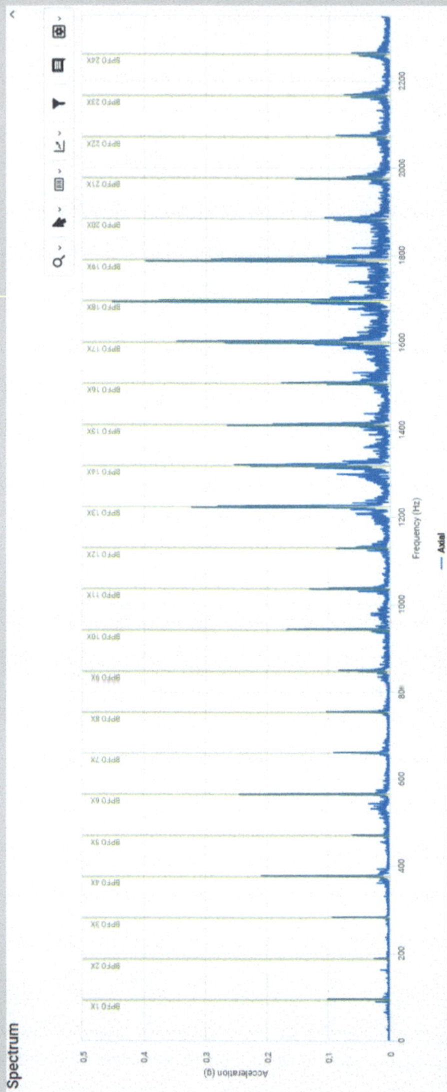

Fig. 6.13 Acceleration-frequency spectrum indicating peaks aligning with the BPFO indicators

Fig. 6.14 Damage to the bearing

Clear Upstream is supplied to customers (e.g., water utility companies) as a Service-as-a-Solution (SaaS) package (ibid.). The solution provides the customers with: (i) IoT connected electrochemical and optical sensors that sample wastewater flows automatically; (ii) a cloud-based predictive analytics engine; and (iii) Kando's methodology (e.g., Steutel-Maron,

2019a). Figure 6.15 is depicting Kando staff installing an IoT unit in the sewer manhole.

Fig. 6.15 Kando staff installing an IoT unit in the sewer manhole (Gelman 2020)

The primary purpose of Clear Upstream is to detect, trace and sample in real time non-conforming loads contained in wastewater being discharged to sewer (ibid.). For this purpose, key metrics are estimated such as: industrial wastewater vs. domestic wastewater, wastewater characteristics and contaminant levels (Steward, 2020). After data collection, Clear Upstream analyses are carried out to assess whether the key metrics deviate from their typical ranges (Steutel-Maron, 2019a). When such anomaly has been identified, Clear Upstream sends push notification alerts in real time (Kando, 2020b). It alerts the operator on an interactive dashboard to incidents that might impact on the WWTP, and automatically start taking samples, if automatic samplers have been installed (e.g., Steutel-Maron, 2019a). Finally, the solution identifies the alleged source of the sewer discharge (ibid.).

This allows the operator to react before the contaminants reach the WWTP downstream and impact on the WWTP processes and assets (Steutel-Maron, 2019b). By identifying the source upstream, the operator can take informed decisions (ibid.). The solution enables WWTPs to operate more efficiently by reducing energy and materials consumption, costs and labour demands (Steward, 2020). In parallel, it improves the effectiveness of identifying excessive wastewater contamination and alleged sources of such contamination (ibid.). In parallel, this might also assist the operator with identification of sources of odour, infiltration/inflow and wastewater discharges to stormwater collection systems (ibid.).

All laboratory test results can be added to Kando's database to inform analysis in similar situations and enable improved responses to future sewer discharges (Steutel-Maron 2019a). Contamination *hot spots* and trends can be identified to assist the operator to contain future incidents (Steward, 2020).

Kando (Israel): Case Studies (Goal, Method, Result)

EYDAP Water Utility (Greece)

As reported, the goal was to minimise the inflowing WWTP loads and reduce OpEx, but major illegal industrial discharge sources were unknown (Kando, 2020c).

The system initially installed with the sewer system included four IoT connected sensors and sampling units across EYDAP's network with one pair in proximity of the investigated industrial plant's discharge point (Steutel-Maron, 2019a).

During the discharge event, the IoT connected sampling units activated autonomously and the collected samples were sent for laboratory analyses (ibid.). The laboratory results confirmed the out-of-range contaminant levels, but Kando's algorithm indicated, as a source, a different industrial plant than that was originally suspected (ibid.). The equipment was thus relocated to confirm the source (ibid.).

Following the equipment relocation, Kando's engine immediately detected a contaminated load and correlated the load to a specific location (ibid.). The subject premises were further monitored, with further samples taken automatically during subsequent incidents (ibid.). Laboratory results confirmed the initial Clear Upstream's findings (ibid.). After the verification, EYDAP was able to make informed decisions in relation to the source premises and reduce the subsequent contamination incidents (ibid.).

In summary, the project outcomes were as follows:

- Previously unknown illegal discharges were identified within the first weeks of the project (Kando, 2020c).
- The illegal discharges were ceased what immediately reduced the inflowing load to WWTP (ibid.).
- EYDAP was able to reduce the discharges of contaminated loads by 50% (Steutel-Maron, 2019a).
- 1.3 million USD as *pollution damages* was paid by the offenders (Kando, 2020c).

Berlin (Germany) (Kando, 2020b*)*

In this case study, Kando's Clear Upstream units together with electric conductivity (EC) sensors provided by Berliner Wasserbetriebe (BWB), were used in the central-western part of Berlin to identify illegal, but unintentional cross-connections between sanitary sewers and stormwater collection system. This resulted in raw sewage being discharged to the storm collection system and subsequently to the receiving waters without being treated.

Clear Upstream acted as *a hotspot screening method* for identifying sections of the sewer catchment with a high likelihood for illegal connections. Starting at the discharge point to the storm sewer, sensors were iteratively re-located upstream until a hotspot area was finally identified. In the second step, digital temperature sensors were planned to be applied to all sewers within the hotspot area to identify the accurate locations of such illegal connections.

Smart Water Meters and Smart Water Networks

Smart Water Meters and Smart Water Networks: Technology/Solution

Smart Water Meters

Digital smart meters allow the utility companies a continuous consumption reading, recording, reporting (daily or at more frequent time intervals) and billing (Lloret et al., 2016). They enable two-way real time communication between meters and the company's control system (ibid.). The purpose is to collect (ibid.):

- Demand data
- Outage management (e.g., service interruption and restoration)
- Quality of service monitoring
- Distribution network analysis with distribution planning (e.g., peak demand and demand reduction)

- Customer billing
- Work management.

Digital smart water systems are computerised systems that measure, collect, analyse distribution and consumption impacts in real time within a metered area (e.g., Lloret et al., 2016). They communicate with metering devices according to a predefined schedule (e.g., every hour) or when required (ibid.). This potentially allows for data analytics and balancing within the metered areas to detect/respond to water leaks, thefts and other water losses (ibid.). They can detect minor leaks and other malfunctions before pipe bursts and prevent significant water losses and water supply service disruptions (Voutchkov, 2019). For this purpose, work orders might also be automatically issued (ibid.).

With smart meters, there is an opportunity to interact with customers by providing them with cloud-enabled information based on data collected (Microsoft, 2016). Their benefits include also the following: more accurate billing; better customer service with fewer on-site visits; sufficient data for planning purposes; identification of water losses and thefts; revenues from the previously unaccounted water usage; customers' online access and water use goal setting/exceedance notifications; water and energy conservation/lower OpEx and potential application of Time of Use Tariffs (TOUT) (ABB Wireless, 2017). The purpose of TOUT is to shift a portion of demand from peak hours to other hours of the day or week (Coles, 2011).

Automated Meter Reading (AMR) Systems and Advanced Meter Infrastructure (AMI)

There has been a recent transition from the simplest Automated Meter Reading (AMR) systems to Advanced Meter Infrastructure (AMI) (Lloret et al., 2016). In summary, AMR allows only for unidirectional communication to meters, but AMI bidirectional communication to meters (ABB Wireless, 2017). Additional benefits of AMI systems include a faster response to water leaks and other wastage issues and a possibility to turn on/off water supply service remotely (ibid.).

Smart Water Networks

As an example, Xylem's smart water system combines smart water meters, advanced sensors, software analytics and services with the FlexNet communication network which reportedly goes beyond AMI. FlexNet communication system is a platform delivering high-speed and secure data via long-range radio (IoT) (Xylem, 2020b). It enables smart meters and sensors to securely transmit and receive near real time customer usage data. As reported, FlexNet technology offers a dedicated radio spectrum that is protected by law from interference (ibid). This technology originates from Xylem's brand Sensus.

Xylem's View, LeakView, SurgeView solutions include real time leak detection, NRW reporting, identification of pipe network assets at risk of failure, operational simulations, and water quality monitoring (Xylem, 2018). These solutions originate from Xylem's brand, Visenti.

View is a IoT solution with analytics designed for water networks. It manages data from different sensors and has real time analytical capabilities to monitor, detect and notify about anomalies related to pressure variations, night flow, water quality issues, demand fluctuations and NRW tracking (ibid.). LeakView is a real time leak and burst detection solution with pressure and acoustic sensors, which is synchronised with GPS and provides alerts to customers before leaks become burst (ibid). SurgeView uses high-rate pressure data to identify and predict, which assets (pipes) within a water network are in a poor condition and likely to fail (ibid). This allows preventative maintenance to be undertaken to extend pipe life (ibid.).

Optimising Leak Detection Using Virtual District Metering Areas (vDMAs) (Xylem, 2020c*)*

The practice of dividing water distribution networks into sectors called District Metering Areas (DMAs) has focused on reducing water loss and good NRW management. At present, the use of digital technologies to create vDMAs eliminates the limitations of the physical DMA by utilising data to provide actionable insights into the condition of their networks. This assists with the automation of the water loss process. A pipe leak will typically lead to a noticeable change in the flow velocity and pressure in the surrounding area. This measurable information can be used to locate the leaking pipe.

DMAs and vDMAs allow operators to use flow meters to closely monitor flow into these areas and apply night-time low-flow monitoring techniques to identify leaks. Unlike a traditional DMA, the monitored area of a vDMA has no strict hydraulic boundaries and does not require actual boundary valve closing and pipe isolation, eliminating hydraulic performance or water quality issues.

Smart Water Meters and Smart Water Networks: Case Studies (Goal, Method, Result)

Wide Bay Water (WBW) (Queensland, Australia)

WBW introduced smart metering technology (AMR) across Harvey Bay City in 2006 (e.g., Cole, 2011). It captures hourly use data and this allows to identify water use patterns for every domestic and commercial customer (ibid.). Before 2010, 24 000 smart meters were in place (ibid.). Since 2012, WBW has been in the process of installing Low Power Wide Area Network (LPWAN) taggle devices on their meters (Taggle Systems, 2019). AMR Module Taggle MRC-1 is a low-power transmitter designed for use with Elster's V100 water meter, which is reportedly used in most of the domestic residencies in Australia (ibid.).

Based on the water use patterns, TOUT was considered in 2009/2010 (as an Australian National Water Commission research project) to encourage water efficiency (Cole et al. 2012). TOUT was planned to target high water users (greater than 600 L/hr) and reduce their annual peak hour (peak hourly demand) and peak day demand (maximum demand in any one day of the year), which would create spare capacity within the system (ibid.). TOUT was researched with various tariff designs at that time, which would target the high outdoor usage considered as having a greater price elasticity in Australia than the indoor usage (ibid.).

The project demonstrated that the traditional TOUT was not considered suitable in the local circumstances and the WBM tariff was subsequently designed on the basis of the existing two-component tariff with a volumetric component, that is, a penalty charge for all consumption greater than 600 L/h (ibid.).

PUB (Singapore)

PUB has reportedly adopted an AMI system for their entire water distribution network including its design, operation and maintenance for the water supply continuity (Voutchkov, 2019). The key PUB objectives include improved: (i) asset management; (ii) water conservation; and (iii) customer service (ibid.).

The AMI system provides continuous monitoring and domestic water consumption data collection with residents being informed on their water usage patterns and alerts them to potential water leaks and excessive water consumption (ibid.). In addition, PUB remotely monitors water consumption of a number of commercial/industrial customers and plans to identify relevant water efficiency goals and good practice guidelines for various industry sectors (ibid.). In addition, PUB is reportedly planning to deploy sensors for faster and more accurate detection of contaminants, and better data analytics to exclude false alerts (ibid.).

This also involves more than hundreds of wireless sensors installed within the water network to collect real time pressure, flow and water quality data. PUB's risk assessment procedures and predictive software tools assist with identification of the top 2% of high-risk pipes for the annual replacement (ibid.). Transmission pipes are being monitored online with the accuracy of 10 m and the operator being alerted within 24 hours of any leak detection (ibid.).

Xylem (United States) (Xylem, 2020c)
Xylem provided assistance to the United Arab Emirates' water authority that was reportedly experiencing a significant water loss in its water distribution network with no evident root causes. For this process, ten vDMAs were created covering 200 miles of pipes (see Fig. 6.16).

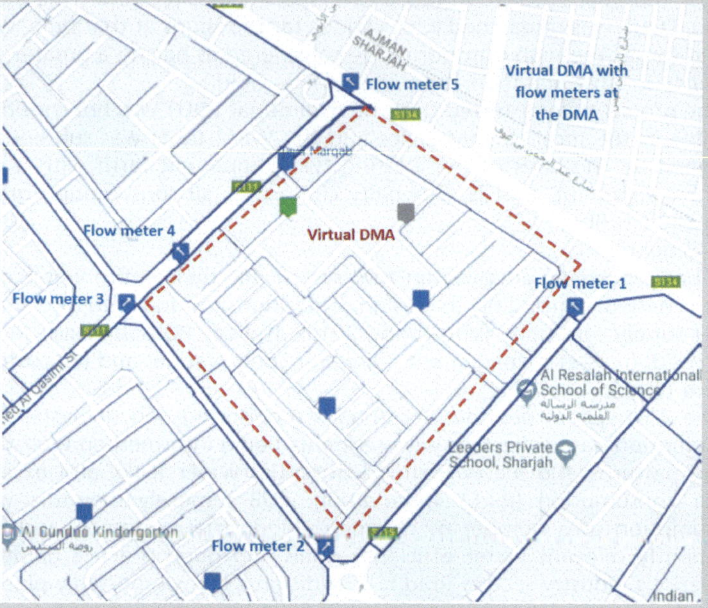

Fig. 6.16 An illustration of vDMAs developed by Xylem for a water authority in the United Arab Emirates

The vDMA implementation included installation of inline flow meters and real time monitoring and analytics to calculate water loss for each vDMA. In addition, pressure acoustic sensors were installed to detect leaks and pressure transients that damaged the pipes. As a result, the water authority could visualise the source of water losses and identify the specific vDMAs where AMI would effectively reduce the losses.

References

Adams, M. L. (2009, December 23). *Rotating machinery vibration: From analysis to troubleshooting* (2nd ed.). CRC Press.

ABB Wireless. (2017). *Application data sheet automated water meter reading.* https://library.e.abb.com/public/95ed641962394e778ed022adfc9dbb58/automated-water-meter-reading.pdf. Accessed 8 July 2020.

Bakker, M., Vreeburg, J. H. G., Palmen, L. J., Sperber, V., Bakker, G., & Rietveld, L. C. (2013). Better water quality and higher energy efficiency by using model predictive flow control at water supply systems. *Journal of Water Supply: Research and Technology—AQUA, 62* (1), 1–13.

Cole, G. (2011, October). *Time of use tariffs: Reforming the economics of urban water supply, waterlines report*. National Water Commission.

Cole, G., O'Halloran, K., & Stewart, R. A. (2012, January 12). Time of use tariffs: Implications for water efficiency. *Water Supply*, 12 (1), 90–100. https://iwaponline.com/ws/article-pdf/12/1/90/678710/90.pdf. Accessed 24 July 2020.

E+H. (n.d.). *Predykcyjny system sterowania Liquiline Control CDC81.* Endress + Hauser. https://www.pl.endress.com/pl/wieksza-wydajnosc-mniejsze-koszty/rozwiazania-analityczne/usuwanie-azotu/kontrola-napowietrzania-usuwanie-fosforu-w-oczyszczalni-sciekow. Accessed 20 July 2020.

E+H. (2018a). *Controlled by Liquiline Control Automated nitrogen removal in Stadtlohn wastewater treatment plant*. Endress + Hauser case study sheet. Endress + Hauser. https://www.pl.endress.com/pl/Endress-Hauser-Twoj-Partner/nasze-realizacje/stadthol-usuwanie-azotu. Accessed 21 July 2020.

E+H. (2018b). *Controlled by Liquiline Control. Automated phosphate removal in Stadtlohn wastewater treatment plant*. Endress + Hauser case study sheet. Endress + Hauser. https://www.pl.endress.com/pl/Endress-Hauser-Twoj-Partner/nasze-realizacje/automatyzacja_stracania_fosforu. Accessed 21 July 2020.

E+H. (2020). Jak zwiększyć efektywność i oszczędzać na kosztach oczyszczalni ścieków? Wywiad ze specjalista ZGWK w Tomaszowie Mazowieckim. *Kurier Wod-Kan. Magazyn klientów Endress + Hauser* (pp. 24–28).

Gelman, J. (2020). *Personal communication.* Kando. 1 October 2020.

Kando. (2020a, June 15). *Kando and Paperspace Partner to Bring Advanced Machine Learning to Municipal Systems Monitoring.* https://blog.paperspace.com/kando-and-paperspace-partner-to-bring-advanced-machine-learning-to-municipal-systems-monitoring/. Accessed 22 July 2020.

Kando. (2020b). *Sensors and smart analytics for tracking illicit sewer connections hotspots.* https://www.digital-water.city/solution/sensors-and-smart-analytics-for-tracking-illicit-sewer-connections-hotspots/. Accessed 22 July 2020.

Kando (2020c, June). *Kando Customer Success Stories.*

Lipnicki, P. (2021, January 5). *Personal communication.* ReliaSol.

Lloret, J., TomásGironés, J., Canovas Solbes, A., & Parra-Boronat, L. (2016). An integrated IoT architecture for smart metering. *IEEE Communications Magazine, 54*(12), 50–57. Institute of Electrical and Electronics Engineers (IEEE).

Lubbers, C. L., Icke, O., van Eijden, R., de Koning, M., Huising C., & de Wit R. (2018, July 17–18). *Performance improvement of wastewater transport systems and treatment processes by performance monitoring and predictive control.* In the proceedings of 12th European Waste Water Management Conference, Aqua Environment, Manchester, United Kingdom.

Mechefske, C. K. (2005). Machine condition monitoring and fault diagnosis, Chapter 25. In C. W. de Silva (Ed.), *Vibration and Shock Handbook.* CRC Press, Taylor and Francis Group.

Mitchell, J. S. (1978, December 10–15). *Bearing diagnostics: An overview* (pp. 15–24). Winter Annual Meeting. American Society of Mechanical Engineers.

Microsoft. (2016, October). *Water industry: Benefits of moving to Cloud technology.* Prepared for Water Sector Organisations, United Kingdom. https://info.microsoft.com/rs/157-GQE-382/images/WaterSectorCloudPoV.PDF. Accessed 8 July 2020.

Passos, C. (2020, July 16). *Personal communication.* Dynamox.

Pfeiffer, S. (2017). The vision of "industrie 4.0" in the making—A case of future told, tamed, and traded. *Nanoethics, 11*, 107–121. Springer Nature B.V. Published online on 25 January 2017. https://link.springer.com/article/10.1007/s11569-016-0280-3. Accessed 19 August 2020.

Piotrowski, J. (2007). *Shaft alignment handbook* (3rd ed.). CRC Press, a Taylor & Francis Group.

RHDHV. (2020a). Aquasuite. White paper: *Introducing* the water industry's virtual *operator which never sleeps. Freeing up your operational team's time for high value tasks.* Royal HaskoningDHV. https://analytics-eu.clickdimensions.com/royalhaskoningdhvcom-agyu4/pages/m0sryp6feeqiwbqvrccvq.html?utm_source=Water-for-Industry-securing-business-continuity&utm_medium=AI-POWERED-VIRTUAL-OPERATOR&utm_campaign=Water-for-Industry. Accessed 23 July 2020.

RHDHV. (2020b, January/February). *Aquasuite PURE.* Royal HaskoningDHV. https://aquasuite.ai/en/products/pure/. Accessed 23 July 2020.

Smart Water Magazine. (2020). Technology against climate change: ACCIONA's solutions for more efficient management. *Smart Water Magazine*, SWM Monthly 1. https://smartwatermagazine.com/news/acciona/technology-against-climate-change-accionas-solutions-more-efficient-management. Accessed 10 July 2020.

Steutel-Maron, A. (2019a). *How Big Data enables smart collection systems and protects Wastewater Treatment Plants.* 2 December 2019. https://medium.com/datadriveninvestor/how-big-data-enables-smart-collection-systems-and-protect-wastewater-treatment-plants-32595fbe6852. Accessed 22 July 2020.

Steutel-Maron, A. (2019b). *How to digitize your wastewater network with a smart water solution?* 17 January 2019. https://medium.com/@annelistesutel/how-to-digitize-your-wastewater-network-with-a-smart-water-solution-1eebffb1b836. Accessed 22 July 2020.

Steward, K. (2020). Real time Wastewater Monitoring Enables Rapid COVID-19 Outbreak Detection. *Industry Insight*. 14 July 2020. https://www.technologynetworks.com/applied-sciences/blog/realtime-wastewater-monitoring-enables-rapid-covid-19-outbreak-detection-337391. Accessed 22 July 2020.

Świerczewska, M. (2020). *Personal communication.* Endress + Hauser. 07 August 2020.

Taggle Systems. (2019).https://taggle.com/. Accessed 19 August 2020.

Voutchkov, N. (2019). II. Disruptive Innovations in the Water Sector. In F. Machado & L. M. Mimmi (Eds.), *The future of water a collection of essays on "disruptive" technologies that may transform the water sector in the next 10 years* (pp. 14–44). Inter-American Development Bank, Water and Sanitation Division.

Xylem. (n.d.). *Harness the power of decision intelligence: Driving performance improvement with smarter water,* a brochure. https://www.xylem.com/en-us/campaigns/act/decision-intelligence/. Accessed 10 November 2020.

Xylem. (2018). *Clean water solutions and services: Total control throughout the clean water cycle,* a brochure. https://www.xylem.com/siteassets/industries--applications/resources/xylem-en-uk-clean-water-brochure_en.pdf. Accessed 10 November 2020.

Xylem. (2019). *Cuxhaven, Germany. Reducing aeration energy usage by 26%, minimizing operational expenses, and reducing compliance risk by applying a decision intelligence approach,* a case study leaflet. https://www.xylem.com/en-us/support/case-studies-white-papers/cuxhaven-germany-reducing-aeration-energy-usage-minimizing-operational-expenses-and-reducing-compliance-risk-by-applying-a-decision-intelligence-approach/. Accessed 10 November 2020.

Xylem. (2020a). *Real-time decision support helps cut Combined Sewer Overflow volume by 100 million gallons and reduces capital needed for regulatory compliance. Evansville Water and Sewer Utility, Evansville, IL, USA.* https://www.xylem.com/en-us/support/case-studies-white-papers/real-time-decision-support-helps-cut-combined-sewer-overflow-volume-by-100-million-gallons-and-reduces-capital-needed-for-regulatory-compliance/. Accessed 10 November 2020.

Xylem. (2020b). https://sensus.com/communication-networks/sensus-technologies/flexnet-international/. Sensus, a Xylem brand. Accessed 10 November 2020.

Xylem. (2020c). *White paper: Optimizing leak detection using virtual district metering areas (DMAs).* https://www.xylem.com/en-us/support/case-studies-white-papers/optimizing-leak-detection-with-a-virtual-dma/. Accessed 10 November 2020.

7

A System for Managing Assets Throughout Their Life

Abstract This chapter expands on the initially proposed Water 4.0 adaptation through the companies' ISO 55001 based Asset Management System to approach the climate resilience goal. The system's scope, implementation and benefits are only briefly discussed with the chapter specifically focusing on *digital twins* including Building Information Modelling; alignment of financial and non-financial functions; two approaches to Water 4.0 adaptation through Asset Management System (i.e., adaptability and continuous improvement); and finally, the relevant information security systems. For these processes, the ISO 19650 series of standards supplements the ISO 55000 series to specifically assist with the integration and interaction of information from in the interconnected digital and physical environments—Cyber-Physical System. Water 4.0 adaptation in the absence of proper organisational culture and leadership (maturity) should be made risk-wise by taking small incremental steps (Plan-Do-Check-Act cycle), as usual with the remainder of Asset Management System. To adapt to the change quickly, a typical cause-effect relationship: *people—processes—technology* needs to be followed. In a reciprocal process, *digital twins* can ensure, however, a

R. Kijak, *Water Asset Management in Times of Climate Change and Digital Transformation*, Palgrave Studies in Climate Resilient Societies,
https://doi.org/10.1007/978-3-030-79360-9_7

greater interoperability, interdependence and data/information exchange and since has also a positive impact on the organisational culture and maturity. It is suggested in this chapter that ISO 55001 asset management standard with its focus on balancing performance, risks and costs, and asset management fundamentals could manage complexity of Water 4.0 and in a sense, standardise it without compromising its business value. Finally, *security-minded management* that are of paramount importance for Water 4.0, could also be captured in the Asset Management System unless there is a dedicated security management system already in place.

Keyword Water 4.0 · Asset Management System · Climate resilience · *Digital twins* · Building Information Management · Asset management fundamentals · Adaptability · Asset management maturity · Continuous improvement · Plan-Do-Check-Act cycle · Information management · *Security minded management*

It has been proposed earlier in this book (see Chapter 4) that risk accommodation becomes the primary strategy in the years to come to ensure the required level of water supply services with the climate change accelerating. Risk accommodation is strategised together with the parallel and ancillary retreat/avoid strategy involving primarily Water 4.0 technologies and solutions. In the future, protection against risks will become less realistic with its complexity increased with many prescriptive standards and codes becoming obsolete. In this chapter, it is proposed that the complexity of such integrated approach becomes more manageable through the implementation of an ISO 55001 based Asset Management System (AMS) (ISO, 2014b). The relevant conceptual model is depicted in Fig. 7.1.

The AMS approach assists with managing risks and costs in the process of ensuring the optimal asset performance. The AMS-based approach is likely to increase the level of confidence and hence effectiveness and reliability (i.e., OEE) for the managed assets throughout their life. The AMS scope and implementation is dependent on

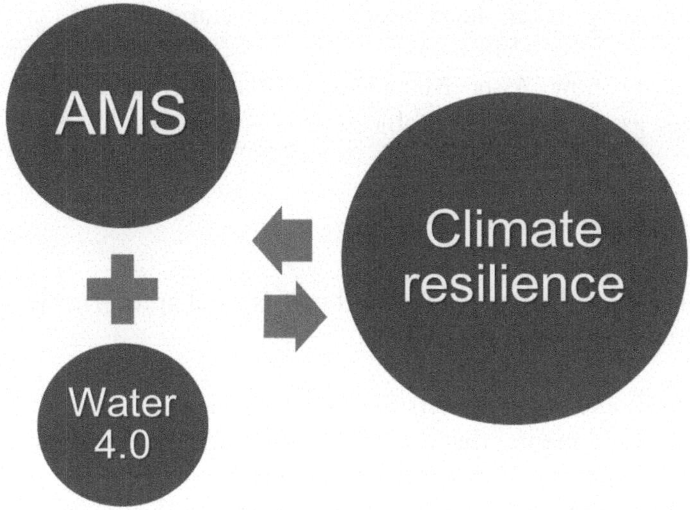

Fig. 7.1 Approaching the climate resilience goal through Water 4.0 incorporated into AMS

asset management fundamentals, that presently include: value, alignment, leadership and assurance (ISO, 2014a). More recently, an additional fundamental is being considered: adaptability (Hardwick et al. 2020), which might be particularly relevant in times of digital transformation (see Sect. 7.4). The business value justifies the existence of assets and would not exist without leadership and organisational culture (ISO, 2018c). Asset management should give assurance for assets to fulfil their purpose (ibid.). Finally, alignment is about translation of organisational objectives into technical decisions and plans (ibid.).

The relevant asset management objectives derived in a bidirectional process from the organisational objectives are normally listed in the key AMS document: Strategic Asset Management Plan (SAMP). SAMP documents the source of such objectives and describes where and how they are applied and achieved. Broadly interpreting ISO 55002 (ISO, 2018c), the purpose of SAMP is as follows:

- Implementing asset management policy's principles

- Documenting stakeholders' needs and expectations (LOS)
- Determining the AMS scope
- Describing how Asset Management Plans (AMPs) relate to the strategic asset management objectives (*line of sight*).

7.1 Scope

Based on earlier discussions in this book (see Chapter 4), the scope of AMS should preferably encompass:

- Both built and natural assets (natural catchments)
- Natural catchments that are delineated on the regional basis (if required based on a prior risk assessment)
- Both water and energy assets to ensure sustainability of energy intense water treatment and transmission technologies (hydro and other forms of energy generated onsite or captured simply as the energy purchase cost).

The knowledge of each asset value is critical for proper asset management practices achieved through condition-based depreciation. (Alignment of financial and non-financial functions is discussed in Sect. 7.4.) This consideration applies also to natural assets. The catchment valuation includes their costs and benefits, for example, avoided/minimised costs of reacting to fire, flood, contaminated runoff, disease outbreaks and other emergencies, if well-managed. Water supply catchments might also have other functions such as recreational, but it is proposed in this book that the valuation of catchments be limited to their primary water supply function.

In some instances, catchment boundaries might be delineated on the regional or even pan-regional basis. This has implications for the AMS boundaries and ownership with potentially multiple owners who must act and cooperate seamlessly to ensure a *line of sight* between the strategic-tactical-operational levels within the system's structure including physical (built/natural) and digital assets. A regional system might look complex, but it is not far different from that how Seqwater is currently managing catchments in southeast Queensland (Australia), which are 95% privately owned. (See the relevant examples from Australia and Malta in Chapters 3 and 4.)

The system might also need to encompass water and energy assets (or consider the purchased energy cost) with, for example, 75% of the cost of water in Malta directly related to energy production (Gourbesville, 2019). As with large-scale desalination plants, their sustainability might only be achieved if renewable energy is used to power them. As this is not always possible, carbon credits might be a minimalistic solution. Such water–energy nexus is also an opportunity for water utility companies to demonstrate sustainability of their practices.

7.2 Implementation

The AMS implementation could be described in eight steps, however, the proposed sequence of initial steps 1–5 is not compulsory as steps are undertaken in an iterative manner. These steps—broadly interpreting ISO 55002 (ISO, 2018c)—are as follows:

1. *Gap analysis*—assessment of the existing business processes against ISO 55001 system requirements
2. Staff training at every stage of the AMS implementation
3. Preparation of an asset management policy, asset management objectives, SAMP and AMPs
4. Draft AMS including the description of all business processes
5. Preparation of an AMS implementation schedule
6. Internal and third-party audits
7. Corrective actions (if needed)
8. Deming's Plan-Do-Check-Act (PDCA) cycle for continual improvement.

The system implementation and potential certification is just the beginning. Therefore, continual improvement based on the PDCA cycle is considered critical for successful system implementation. Figure 7.2 depicts—in a simplified way—how the PDCA cycle applies to the asset life cycle (i.e., asset planning, design, O&M and disposal stages) where

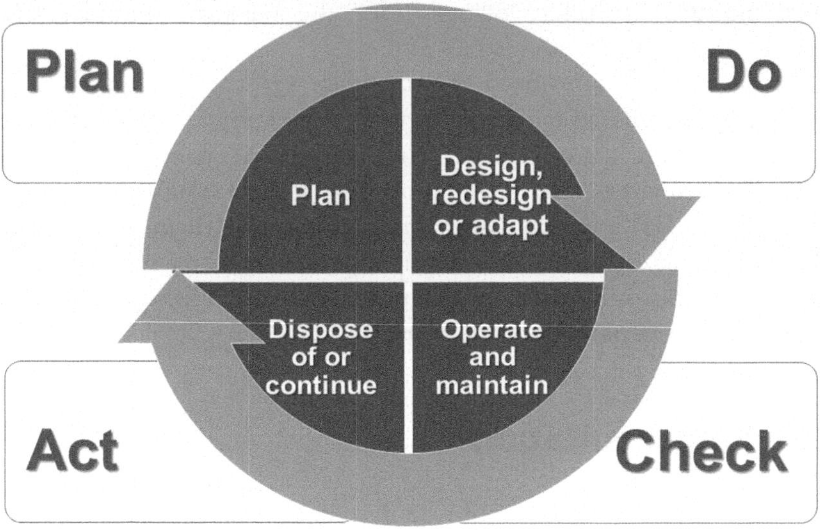

Fig. 7.2 Application of the PDCA cycle throughout the asset life cycle

a decision whether to continue with the operation of an asset or dispose it of can be made as part of Check step of the PDCA cycle. Such a decision can follow a structured process such as Reliability Centred Maintenance (RCM) with an RCM recommendation, for example, that redesign/upgrade of an asset is necessary to prevent its impending functional failure. Even the world class O&M will not improve an obsolete asset design characterised by low OEE and the only solution remains asset redesign or adaptation. Digital technologies and solutions can particularly assist with asset adaptation. With *digital twins* (see Sect. 7.5), assets could be redesigned and adapted (or even operated differently) with greater ease when climate changes.

7.3 How Water 4.0 Fits into an ISO 55001 Asset Management System?

ISO 55000 requires that organisations investigate and acquire knowledge about new asset management technologies and practices, assess their benefits and, if appropriate, incorporate them into their AMS. Asset Management 4.0 and Water 4.0 are not exceptions.

Asset Management 4.0 (see Chapter 5) can support digital transformation within the AMS scope.

These processes are closely aligned with Water 4.0 to ensure efficiency, improve organisational culture and leadership, and maximise the business value. They build upon *evidence-based decision-making* (ISO 2015a) and traditional methods and tools, but with connectivity and use of *big data* and information extracted from *digital twins* or predictive analytics.

7.4 Water 4.0 Implementation Through ISO 55001 Standard

The literature review might suggest that the Water 4.0 concept was deficient and needed an *integrated business model* to succeed (Alabi et al., 2019). Integrated business models would reportedly manage the complexity of respective business processes (ibid). For asset intensive water utility companies that transform their businesses digitally (Water 4.0), the immediate challenges are, however, technological. *Processes* are important, but their improvements require prior or, at least, parallel improvements in organisational culture and leadership (*people*), which might have timelines incompatible with disruptive technologies. Without the proper culture and leadership, business process improvements should be made risk-wise by taking small incremental steps, as with Asset Management Systems being continuously improved in the PDCA cycle.

The problem is not typically with the technology, but with people's *digital mindset* (e.g., Gulati, 2021). The typical cause–effect relationship: *people—processes—technology* needs to be followed. When people do not catch up with *technology*, the organisational culture needs to be improved in the first place, which is a difficult task. To improve processes, a solution is often easier and might involve, for example, engagement of engineering consultants (or autonomous AI-based solutions in the future).

When some authors suggest that Water 4.0 needs an *integrated business model*, others argue that its implementation is slowed down by lack of standardisation (e.g., GWP, 2019; Voutchkov, 2019). As with the integrated business model, standardisation is driven primarily by *people* who might arguably hamper digital transformation rather than lack of standardisation itself. Moreover, digital transformation is a dynamic process with new digital technologies and solutions continuously emerging (Ziemer & Volker, 2017). For such dynamic process that has even been named as *water revolution,* standardisation of Water 4.0 (technologies and respective processes) might be impractical and, in fact, have an opposite and negative outcome.

As with all innovations, Water 4.0 implementation timelines are non-linear so there might be further issues with standardisation. To keep up with the required tools, organisations must innovate, but also learn agility and adaptability, but this requires clearly asset management maturity (including digital maturity), leadership and organisational culture. Asset management maturity involves leadership and a sustainable management of organisational performance through realising value from assets, but is also supported by technical processes (Hardwick et al., 2020). The latter can include in-house digital skills and infrastructure.

Though the potential Water 4.0 standardisation might reduce risks, it would not allow to fully explore opportunities to extract value from assets for the mature organisations (see the previous Chapters 4–6). ISO 55001 asset management standard with its focus on balancing performance, risks and costs, and asset management fundamentals could arguably manage complexity of Water 4.0 and in a sense, standardise it without compromising its business value.

7.5 Digital Twins Transforming Asset Management

There should be no doubt that the application of *digital twins* transforms asset management practices and procedures. As with relevant case studies presented in Chapter 6, opportunities and risks identified in the digital environment could be considered in the physical environment (as with CPS in general). CPS with sensors controlling the physical environment is able to work in real time to influence the physical environment outcomes (ISO, 2020b). A *digital twin* is often linked to an AI-based simulation engine and data analytics (e.g., IET and Atkins).

The process of integration and interaction of information resulting from the digital and physical environments has been depicted in Fig. 7.3 that originates from ISO 19650 (ISO, 2018f). In this instance, it has generally been limited to BIM defined as a shared digital representation of *built assets*. It applies to their life cycle stages (planning, design and engineering, construction, O&M, renewal and end-of-life) (ibid.). Other parts of the series, ISO 19650–2 (ISO, 2018g) and ISO 19650–3 (ISO, 2020a) provide more detailed information on asset delivery and operational phases respectively.

The provision of information is considered as one of the key criteria for the completion of a project (Project Information Management, PIM) or an asset management activity (Asset Information Management, AIM) (ISO, 2018f). The ISO 19650 standard stresses the importance of continuity of information management over the life cycle of built assets (ibid.).

ISO 19650 series supplements ISO 55001 (ISO, 2014b) and contains additional requirements for the AMS development by focusing on asset information. In addition to the necessity for appropriate and timely asset information, two other principles of ISO 55001 are important here: (i) managing assets with objectives originating from asset management policies, strategies and plans (*line of sight*); and (ii) leadership and governance from *top management*. They are part of digital horizontal and vertical integration within an organisation. The ISO 19650 implementation requires a number of plans including a BIM execution plan that

Fig. 7.3 Integrated and life cycle organisational management with ISO 9001 (ISO, 2015b), asset management with ISO 55001 (ISO, 2014a) and information management with ISO 19650 (ISO, 2018f) within the physical and digital environments. The original figure 3 from ISO 19650–1: 2018: Organization and digitization of information about buildings and civil engineering works, including building information modelling (BIM)—Information management using building information modelling—Part 1: Concepts and principles, is reproduced with the permission of the International Organization for Standardization, ISO. This standard can be obtained from any ISO member and from the website of the ISO Central Secretariat at the following address: http://www.iso.org. Copyright remains with ISO

determines the delivery team's capability and capacity in the information management process (ISO, 2018f).

As with hydraulic models operationalised through real time data in case studies discussed in Chapter 6, BIM represents the initial maturity level for *digital twins*. At a more advanced maturity level, BIM might be supplemented, for example, with wireless sensor data (IET and Atkins). BIM requires the creation of a *digital twin* of a proposed asset (structure, plant or facility). As a result, planning, construction and O&M processes can be managed with ease and various engineering disciplines closely collaborating throughout these life cycle stages (e.g., GWP, 2019). This is to ensure a greater interoperability (interdependence) and data/information exchange and since has also a reciprocal and positive impact on the organisational culture and maturity.

In general, BIM enables to construct and subsequently manage new assets digitally with fewer on-site safety and environmental impacts, errors and discontinuities, that is with the business value improved throughout their life cycle (IET and Atkins). It can also support the retrofit of existing built assets (ibid.). A *single version of truth* becomes available for assets where all relevant data are accessible throughout their life cycle (ibid).

Beyond BIM, there is likely to be a two-way data integration and interaction, and finally a fully autonomous O&M (IET and Atkins). Beyond predictive or even prescriptive analytics, there is AI-based self-governance and transformative machine-to-machine analytics that would allow asset engineers to have more conceptual and strategic roles (see Chapter 4). Machines would talk directly to fellow machines when arduous tasks are being done.

7.6 Alignment of Financial and Non-Financial Functions

To demonstrate the achievement of organisational objectives associated with Water 4.0, technical asset information should be aligned with the corresponding financial information from the finance and accounting departments of water utility companies. This is broadly consistent with the ISO 55001 standard requirements (ISO, 2014b) and more recent standard ISO 55010 on aligning financial and non-financial functions in asset management (ISO, 2019a).

In accordance with ISO 55010, the integration is both vertical and horizontal (ISO, 2019a). Vertically, it involves *top management* financial/non-financial directives (top-down) and with the bottom-up feedback loop (ibid.). Horizontally, there are information flows between organisational departments (ibid.). Technical and financial managers are jointly responsible for delivering asset management, but their departments have historically developed their own practices (ibid.). The goal is to remove organisational silos to jointly pursue common organisational objectives (ibid.). In addition to the financial objectives, mutually beneficial non-financial objectives can be identified to extract a greater business value from assets (ibid.).

To provide the supporting financial information, ISO 55010 specifically recommends estimation of TotEx (Total Expenditures) to ensure sustainable CapEx and OpEx decisions throughout the asset life cycle (ibid.). TotEx is a sum of CapEx and OpEx over a defined period (ibid.). For this purpose, condition-based depreciation might be prudent to maintain the asset value in this period.

As with other asset life cycle stages (see Sect. 7.2), O&M can become more effective when the asset value has been predetermined and O&M departments have been allocated a well-defined portion of the overall capital and operational budgets.

7.7 Adaptability or Continuous Improvement

Adaptability has recently been considered by Hardwick et al. (2020) as more agile and responsive than continuous improvement (see Glossary). It has been suggested that adaptability supplements other asset management fundamentals included in ISO 55000 (ISO, 2014a) and supports the asset management maturity (Hardwick et al., 2020). (The current asset management fundamentals and asset management maturity are discussed in more detail at the beginning of this chapter).

Without integrated business models and standardisation (see Sect. 7.4), adaptability associated with Water 4.0 seems to be a preferred approach, which also brings a greater climate resilience faster (*big bang*). This requires collaborative culture (interdependence) and asset management maturity, but promotes innovation and brings the greatest business value (ibid.). Adaptability allows to progress faster and introduces the required change (e.g., new design and operational guidelines and digital transformation). Such change requires a *digital mindset* and a different organisational culture (Gulati, 2021). "Progress is a nice word. But the change is its motivator. And change has its enemies" Kennedy (1964).

It is then prudent that every organisation sets their own standards and objectives through their own AMS, which consider their preferred level of adaptability (change). Adaptability is arguably required for Water 4.0 with adaptation bringing the greatest value (Weitze et al., 2018), see Fig. 7.4. Consequently, it also brings climate adaptation.

As earlier discussed in Chapter 4 and demonstrated through case studies in Chapter 6, adaptability seems to be an agile and prudent approach during the *fourth water revolution*, but requires a sufficient level of asset management maturity (Hardwick et al., 2020). In this

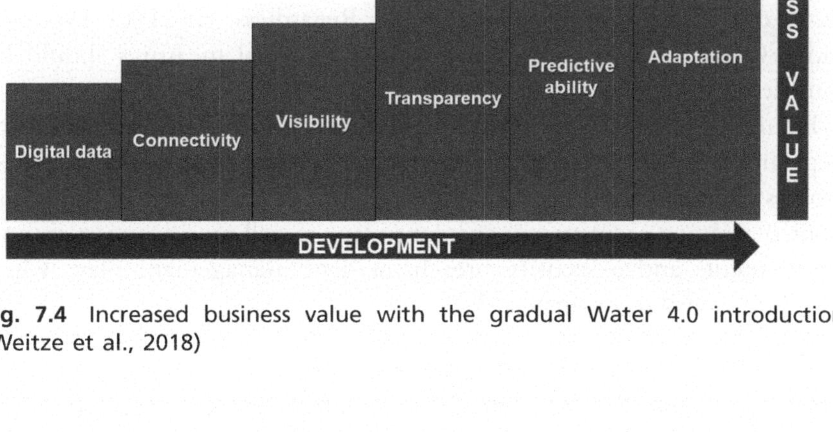

Fig. 7.4 Increased business value with the gradual Water 4.0 introduction (Weitze et al., 2018)

instance, the asset management maturity includes digital culture. Alternatively, Water 4.0 can also be retrofitted in a step approach as suggested by Greenfield (2020). The entire system overhaul is not the only path to digital transformation (ibid.). The alternative path can include, for example, adding only connectivity with wireless sensors to the existing machines (e.g., accelerometers attached to pumps) allowing them to be remotely monitored over the internet (ibid.). Greenfield (2020) suggests that such step approach is less risky and less costly, and consequently more consistent with the definition of continuous improvement.

7.8 Strategising Information Security

Cybersecurity extends clearly beyond technical issues that have been discussed earlier in Chapter 5. A strategic approach is then recommended with the evaluation of all opportunities and risks for all processes and assets managed by a water utility company (e.g., GWP, 2019). To capture information security issues in the AMS, ISO 19650–5 (ISO, 2020b) assists with the relevant requirements on *security-minded management*. The information security issues could be addressed through SAMP or a separate strategy, in either case with subordinate plans at the tactical and operational levels (*line of sight*). If an organisation has already an Information Security Management System (ISMS) developed in accordance with ISO/IEC 27000 (ISO, 2018h), then ISMS, AMS and other management systems might be integrated (see Sect. 7.9). Such integration is, in fact, recommended by ISO 55002 (ISO, 2018c) to avoid duplications of work and data. Regardless, the ISO 19650–5 and ISO/IEC 27000 technical and organisational measures should be considered in parallel (ISO, 2020b).

Figure 7.5 contains the relevant flowchart followed by explanation of steps and keys in accordance with ISO 19650–5 (ibid.). The depicted process includes, amongst other things, the development of an initial security strategy, security management plan, as well specific measures and requirements and a security breach/incident management plan, when and if necessary to mitigate risks.

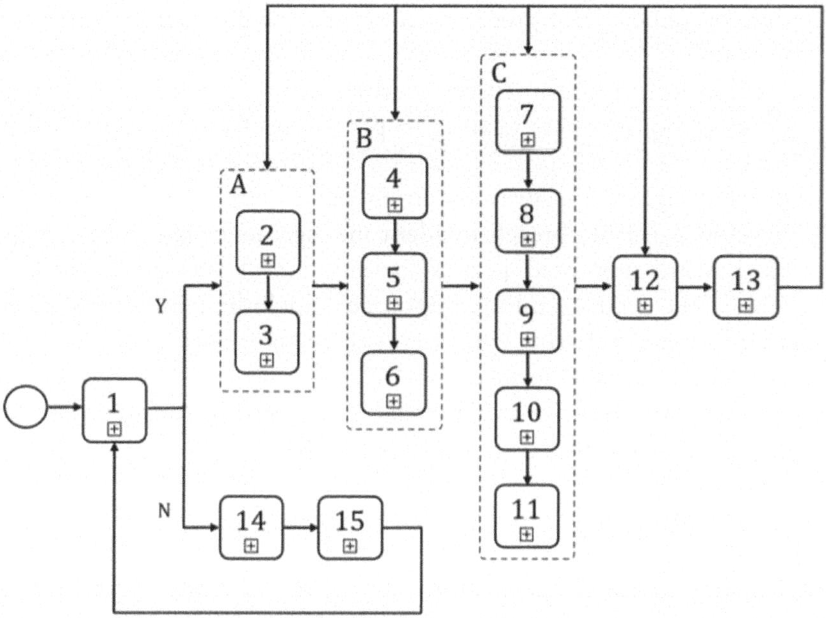

Fig. 7.5 Security-minded management in accordance with ISO 19650–5 (ISO, 2020b)

Where

A initiate a security-minded approach
B develop a security strategy
C develop a security management plan
Y yes
N no
1 determine, using the security triage process whether a security-minded approach is required
2 establish governance, accountability and responsibility arrangements for the security-minded approach
3 commence development of the security-minded approach
4 assess the security risks
5 develop security mitigation measures
6 document tolerated security risks

7 develop policies and processes to implement the security mitigation measures
8 develop security information requirements
9 develop requirements relating to provision of information to third parties
10 develop logistical security requirements
11 develop a security breach/incident management plan
12 work with appointed parties in and out of formal contracts to embed the security-minded approach, including the development of information sharing agreements where necessary
13 monitor, audit and review
14 protect any sensitive commercial and personal information (no other security-minded approach required)
15 review if there is change in the initiative, project, asset, product or service which may impact on its sensitivity

7.9 Other Opportunities of the AMS Development

The AMS development provides high-level opportunities associated with asset management. They include: (i) improved financial performance; (ii) informed asset investment decisions; (iii) managed risk; (iv) improved services and outputs; (v) demonstrated social responsibility; (vi) demonstrated compliance; (vii) enhanced reputation; (viii) improved organisational sustainability; and (ix) improved efficiency and effectiveness (OEE) (ISO, 2018c). There are also more specific opportunities for the AMS development as reported in journal articles or from the author's experience. In addition to those associated with climate change and digital transformation discussed throughout this book, the following opportunities of the AMS implementation have also been identified:

- Ease of integration with other organisational management systems such as HACCP or ISO 22000 (ISO, 2018a), ISO 9001 (ISO, 2015b), ISO 14000 (ISO, 2015c), ISO 45001 (ISO, 2018d) and ISO/IEC 27000 (ISO, 2018h)
- More favourable court's decisions in class actions (Kennedy, 2017)
- Justification of maintenance budgets to the owners/boards (ibid.)
- As assets are managed by balancing their cost, risk and performance, providing grounds for proposed water charges (tariffs) to the regulator, for example, Independent Pricing & Regulatory Tribunal (IPART) in the Australian state of New South Wales or Queensland Competition Authority (QCA) in the state of Queensland (BSI, n.d.)
- Reduction of capital expenditures and operational costs by 20 and 10 percent respectively and a greater level of sustainability and compliance with environmental regulations (Williams, 2017)
- Better risk management, a transparent prioritisation of capital projects, documenting business processes and procedures, conscious planning of training and competence retention, a better cooperation of the company's organisational units without creating *silos* (ibid.).

References

Alabi, M. O., Telukdarie, A., & Janse van Rensburg, N. (2019, December). *Water 4.0: An integrated business model from an industry 4.0 approach.* In The proceedings of the 2019 IEEE International Conference on Industrial Engineering and Engineering Management (IEEM). https://www.researchgate.net/publication/339021634_Water_40_An_Integrated_Business_Model_from_an_Industry_40_Approach. Accessed 8 July 2020.

GWP. (2019). *Water 4.0. Made in Germany.* German Water Partnership.
Glickman, J., & Leroi, A. (2015). *Adapt and adopt: Digital transformation for utilities.* Bain & Company. https://www.bain.com/insights/adapt-and-adopt-digital-transformation-for-utilities/. Accessed 8 June 2020.
Gourbesville P. (2019). Smart water solutions for water security: From concept to operational implementation. In K. Lim, A. K. Makarigakis, O. Sohn, & B. Lee [(Editor-in-Chief)], *Water security and the sustainable development goals* (pp. 47–67). Global water security issues series. United Nations Educational, Scientific and Cultural Organization (UNESCO) International Centre for Water Security and Sustainable Management. https://unesdoc.unesco.org/ark:/48223/pf0000367904.locale=en. Accessed 18 March 2020.
Greenfield, D. (2020). A step approach to industry 4.0. *Automation World.* 04 June 2020. https://www.automationworld.com/factory/iiot/article/21135339/a-step-approach-to-industry-40. Accessed 18 June 2020.
Gulati, R. (2021). *Maintenance and reliability best practices* (3rd ed.). Industrial Press.
Hadjimichael, A., Comas, J., & Corominas, L. l. (2016, December 1). *Do artificial intelligence methods enhance the potential of decision support systems?* A review for the urban water sector. To appear in: AI Communications.
Hardwick, J., Killeen, M., Kohler, P., Lafraia, J., & Nugent, S. (2020, April 2). *Living asset management.* Living Asset Management Think Tank Incorporated.
ISO. (2014a). *ISO 55000:2014 Asset management—Overview, principles and terminology.* International Organization for Standardization.
ISO. (2014b). *ISO 55001:2014 Asset management: Management systems—Requirements.* International Organization for Standardization.
ISO. (2015b). *ISO 9001:2015 Quality management systems—Quality management systems—Requirements.* International Organization for Standardization.
ISO. (2015c). *ISO 14001:2015 Environmental management systems—Requirements with guidance for use.* International Organization for Standardization.
ISO. (2018a). *ISO 22000:2018 Food safety management systems—Requirements for any organization in the food chain.* International Organization for Standardization.
ISO. (2018c). *ISO 55002:2018 Asset management—Management systems—Guidelines for the application of ISO 55001.* International Organization for Standardization.
ISO. (2018d). *ISO 45001:2018 Occupational health and safety management systems—Requirements with guidance for use.* International Organization for Standardization.

ISO. (2018f). *ISO 19650–1: 2018 Organization and digitization of information about buildings and civil engineering works, including building information modelling (BIM): Information management using building information modelling—Part 1: Concepts and principles.* Geneva, Switzerland: International Organization for Standardization.

ISO. (2018g). *ISO 19650–2:2018 Organization and digitization of information about buildings and civil engineering works, including building information modelling (BIM)—Information management using building information modelling—Part 2: Delivery phase of the assets.* International Organization for Standardization.

ISO. (2018h). *ISO/IEC 27000:2018 Information technology—Security techniques—Information security management systems—Overview and vocabulary.* International Organization for Standardization.

ISO. (2019a). *ISO/TS 55010:2019 Asset management—Guidance on the alignment of financial and non-financial functions and asset management.* International Organization for Standardization.

ISO. (2020a). *ISO 19650–3:2020: Organization and digitization of information about buildings and civil engineering works, including building information modelling (BIM)—Information management using building information modelling—Part 3: Operational phase of the assets.* International Organization for Standardization.

ISO. (2020b). *ISO 19650–5:2020: Organization and digitization of information about buildings and civil engineering works, including building information modelling (BIM)—Information management using building information modelling—Part 5: Security-minded approach to information management.* International Organization for Standardization.

Kennedy, J. (2017). Asset management—Defending the business. *The Asset Journal, 11*(3), 4–12. Asset Management Council.

UN. (n.d.). https://www.un.org/sustainabledevelopment/sustainable-development-goals/. United Nations. Accessed 19 August 2020.

Voutchkov, N. (2019). Disruptive Innovations in the Water Sector, II. In F. Machado & L. M. Mimmi (Eds.), *The future of water a collection of essays on "disruptive" technologies that may transform the water sector in the next 10 years* (pp. 14–44). Inter-American Development Bank, Water and Sanitation Division.

Weitze, L., Solas, H., & Palzer, U. (2018). *Digital assessment to increase efficiency in the water infrastructure—Water 4.0.* Weimar Institute of Applied Construction Research. A presentation from German American Dialogue for Sustainable Water Infrastructure and Technologies in Milwaukee/Wisconsin, United States. 18 October 2019. https://www.researchgate.net/publication/328502279_Digital_Assessment_to_Increase_Efficiency_in_the_Water_Infrastructure_-_Water_40. Accessed 9 July 2020.

Williams, W. (2017, November 10). *U.S. utilities weigh the cost and benefit of ISO 55,001 certification. Breaking Energy.* https://breakingenergy.com/. Accessed 29 January in 2020.

Ziemer, C., & Volker, C. (2017, September 1). Water 4.0: What it Means for the German Water Industry. *WaterWorld*. https://www.waterworld.com/international/utilities/article/16201159/water-40-what-it-means-for-the-german-water-industry. Accessed 8 July 2020.

8 Conclusions

Abstract This concluding chapter reiterates major points made throughout this book in the context of climate change and digital transformation. They refer to the book's initial proposition that the water sector's Water 4.0 adaptation increases *sustainability* including a greater climate resilience. It summarises, more strategically, Water 4.0's benefits including those resulting from case studies presented earlier in this book. It draws further and more philosophical conclusions on the water engineers' role in times to come.

Keywords Water 4.0 · Digital transformation · Climate change · Climate resilience · Asset Management System

> Progress is a nice word. But the change is its motivator. And change has its enemies. (Kennedy, 1964)

Due to climate changes accelerating, semi-arid and water stressed countries have become even more exposed and vulnerable. Floods, droughts and fires within natural catchments of the water supply assets have

R. Kijak, *Water Asset Management in Times of Climate Change and Digital Transformation*, Palgrave Studies in Climate Resilient Societies,
https://doi.org/10.1007/978-3-030-79360-9_8

become part of the landscape more than ever. Such events are likely to become more intense and frequent to the extent that some might result in interruptions of essential services and become finally uninsurable. Natural water supply catchments are clearly natural assets that need to be subject to valuation and maintained accordingly. This might arguably include climate considered on the regional basis.

Contextual examples in this book originate from the subtropical region, but most issues raised are perhaps applicable also to the temperate region. The subtropical region has already *learned lessons* and experienced hazards that might occur in the temperate region in the future. *Lessons learned* from the currently water stressed countries together with emerging disruptive technologies could then become part of risk management anywhere in the world.

Water asset engineers' role is, however, limited. They maximise and reduce supply and demand respectively. For this process, built assets need to be considered in parallel with their natural catchments that need to be delineated in some instances on a regional or even pan-regional basis. The supply maximisation should preferably be conducted in a way that does not exacerbate climate change as with the energy intense regional water supply facilities.

Climate resilience based on uncertain projections of climate changes is also highly uncertain. It is then suggested in this book that protecting against climate risks has become an unrealistic task. *If you can't beat them, join them.* The refocus on accommodating risks and retreating/avoiding risks is then a preferred approach, particularly with extensive use of recycled water and smaller (local) desalinisation plants and preferably powered by renewable energy.

The current prescriptive and engineering standards and codes might become obsolete as time passes. Water supply assets might, however, have more frequent upgrades (redesign and adaptation) before their obsolescence. With the unpredictability of weather in the future, the use of scientifically advanced engineering calculus is likely to be required on each occasion. When the scientifically advanced calculus, *digital twinning* and predictive analytics is left to AI, assets could be redesigned and adapted with greater ease when climate changes. An ancillary retreat/avoid strategy involves water supply asset optimisation and consequently minimisation of *waste*. OEE is then maximised and the source water intake reduced.

In this book, management of NRW, CSOs and WWTP bypasses impacting on the water supply directly (*source-to-tap*) and indirectly (*toilet-to-tap*) has been discussed in a context of Water 4.0 's technologies and solutions. (CSO and WWTP bypasses might result, for example, from pump failures and the sewerage system being overloaded.) To prevent both direct and indirect impacts, it might be prudent to embrace Water 4.0 technologies and solutions. There are also additional benefits. With the AI assisted tools, asset engineers will have a more conceptual and strategic role. Fewer *hurdles* in the absorption of disruptive digital tools are then expected in the future.

The water sector has traditionally been conservative in implementing innovative and disruptive technologies, but this seems to be changing. The sector's Water 4.0 adaptation will arguably increase *sustainability* including a greater climate resilience. This has become a proposition in this book, which was examined based on literature and case studies.

Strategic Water 4.0 opportunities identified in literature could include:

- Addressing water demand by an increased water supply (climate resilience)
- Improving environmental performance (including a lower carbon footprint)
- Potential for water supply decentralisation
- Reducing energy consumption and chemical use
- Better operational management and control (process and asset performance)
- Reducing total life cycle costs (TotEx)
- A greater water use efficiency and a lower water footprint
- Improving LOS and customer experience
- Addressing the United Nations SDGs 6—Clean water and sanitation and 13—Climate action.

Case studies are important because they assist with communicating Water 4.0 related opportunities to decision-makers. In this book, case studies have originated from Europe, Americas, Middle East, Southeast Asia and Australia and involved a number of service providers and water utility companies. They are categorised technology-wise as follows:

1. *Digital twins*, SCADA connected sensors, hybrid analytics for process control (German Endress + Hauser)
2. *Digital twins*, SCADA connected sensors, predictive analytics for process control (Dutch Royal HaskoningDHV, American Xylem and Polish ReliaSol)
3. IoT connected sensors, cloud-based hybrid analytics for predictive maintenance (Brazilian Dynamox)
4. IoT connected sensors, cloud-based predictive analytics for sewer discharge identification (Israeli Kando)
5. Smart water meters and smart water networks (Australian Wide Bay Water, Singaporean Public Utilities Board and Xylem for the United Arab Emirates water authority).

The case studies have generally confirmed the prior literature findings. In addition to the reduction of NRW, CSO and WWTP bypasses, Water 4.0 technologies and solutions ensured the optimisation of energy and chemicals consumption, water use efficiency, water quality, WWTP effluent standards and fewer asset breakdowns. Total Expenditures, carbon and water footprints seem to be minimised in parallel. Water 4.0 has a potential for decentralisation, for example through Water 4.0 assisted WWTP effluent reuse.

It has been suggested in this book that disruptive Water 4.0 technologies be incorporated into the companies' Asset Management System (ISO, 2014) to approach the climate resilience goal. For this process, two approaches are possible: adaptability (see Hardwick et al., 2020) and continuous improvement (see ISO, 2015). The former approach is considered in this book as preferred, but requiring a sufficient asset management maturity. Adaptability allows to progress faster and introduce the required change (e.g., new design and operational guidelines and digital transformation). Without proper organisational culture and leadership, business process improvements should, however, be made risk-wise by taking small incremental steps, as with Asset Management

Systems being continuously improved in the PDCA cycle. To adapt to the change quickly, a typical cause-effect relationship: *people—processes—technology* needs to be followed.

In a reciprocal process, *digital twins* can ensure, however, a greater interoperability, interdependence and data/information exchange and since have also a positive impact on the organisational culture and maturity. The ISO 19650 series of standards supplements the ISO 55000 series to specifically assist with the integration and interaction of information from the interconnected digital and physical environments—Cyber-Physical System. Provided that cybersecurity risks are managed, the respective business value, sustainability and ultimately climate resilience have been demonstrated in this book in a qualitative way.

It is finally noted that the above quote from a speech by Robert F. Kennedy has been used in this book to describe the usual roadblocks encountered managing change for climate resilience (e.g., new design and operational guidelines, and digital transformation). Lateral thinkers may perhaps apply it more literally, that is, to anthropogenic climate change driving innovations and those who deny the existence of such change. Such discussion could even reach the highest level of *argumentum ad hominem.* The hope is Artificial Intelligence that is likely to protect the climate better than humans alone (Lovelock, 2019).

References

Hardwick, J., Killeen, M., Kohler, P., Lafraia, J., & Nugent, S. (2020, April 2). *Living asset management maturity*. Living Asset Management Think Tank Incorporated.

ISO. (2014). *ISO 55001:2014 Asset management — Management systems — Requirements.* International Geneva, Switzerland: International Organization for Standardization.

ISO. (2015). *ISO 9001:2015 Quality management systems — Quality management systems — Requirements.* Geneva, Switzerland: International Organization for Standardization.

Kennedy, R. F. (1964, May 25). *Address by Attorney General Robert F. Kennedy before the United States Conference of Mayors, New York Hilton Hotel, New York, 2pm.* Washington, DC: U.S. Department of Justice. https://www.justice.gov/ag/rfkspeeches/1964/05-25-1964.pdf. Accessed 26 December 2020.
Lovelock, J. (2019, July 4). *Novacene: The coming age of hyperintelligence* (1st ed.). Penguin Books Limited.

References

Adams, M. L. (2009, December 23). *Rotating machinery vibration: From analysis to troubleshooting* (2nd ed.). CRC Press.

ABB Wireless. (2017). *Application data sheet automated water meter reading*. https://library.e.abb.com/public/95ed641962394e778ed022adfc9dbb58/automated-water-meter-reading.pdf. Accessed 8 July 2020.

Alabi, M. O., Telukdarie, A., & Janse van Rensburg, N. (2019, December). *Water 4.0: An integrated business model from an industry 4.0 approach*. In The Proceedings of the 2019 IEEE International Conference on Industrial Engineering and Engineering Management (IEEM). https://www.researchgate.net/publication/339021634_Water_40_An_Integrated_Business_Model_from_an_Industry_40_Approach. Accessed 8 July 2020.

Alanen, P. (2019, February 14). How artificial intelligence is transforming the water sector: Case Ramboll. *Silo.ai*. https://silo.ai/how-artificial-intelligence-is-transforming-the-water-sector-case-ramboll/. Accessed 7 June 2020.

Alegre, H., Baptista, J. M., Cabrera Jr., E., Duarte, P., Merkel, W., & Parena, R. (2006). *Performance indicators for water supply services* (2nd ed.). IWA Publishing. Table 21 (p. 30). © IWA Publishing.

R. Kijak, *Water Asset Management in Times of Climate Change and Digital Transformation*, Palgrave Studies in Climate Resilient Societies,
https://doi.org/10.1007/978-3-030-79360-9

Almagor, D., Lavid D., Nowitz, A., & Vesely, E. (2019). *Maintenance 4.0 implementation handbook*. reliabilityweb.com. Fort Myers, FL, United States.

Bakker, M., Vreeburg, J. H. G., Palmen, L. J., Sperber, V., Bakker, G., & Rietveld, L. C. (2013). Better water quality and higher energy efficiency by using model predictive flow control at water supply systems. *Journal of Water Supply: Research and Technology—AQUA, 62* (1), 1–13.

Ball, J., Babister. M., Nathan, R., Weeks, W., Weinmann, E., Retallick, M., & Testoni I. (Eds.). (2019). *Australian rainfall and runoff: A guide to flood estimation*. Geoscience Australia, © Commonwealth of Australia (Geoscience Australia).

Beckett, R. C., Chapman, R. L., Berendsen, G., Dalrymple J., & Quispe-Chavez, N. (2019). *Quality 4.0' and water management practices*. A Paper for a Presentation at ANZAM 2019 Conference. https://www.researchgate.net/publication/340132763_%27Quality_40%27_and_Water_Management_Practices. Accessed 7 June 2020.

Bilocca. (2018, October 21). Works to start on Gozo reverse osmosis and 'new water' network for south of Malta. *Malta Today*. https://www.maltatoday.com.mt/environment/energy/90293/works_to_start_on_gozo_reverse_osmosis_and_new_water_network_for_south_of_malta#.X6mQBVqSkph. Accessed 15 January 2020.

Blumensaat, F., Leitao, J. P., Ort, C., Rieckermann, J., Scheidegger, A., Vanrolleghem, P. A., & Villez, K. (2019, June 7). How Urban Storm- and Wastewater management prepares for emerging opportunities and threats: Digital transformation, ubiquitous sensing, new data sources, and beyond—A horizon scan. *Environmental Science Technology, 53*, 8488–8498.

BOM. (n.d.). *Drought knowledge centre: Understanding drought*. http://www.bom.gov.au/climate/drought/knowledge-centre/understanding.shtml. Accessed 8 January 2021.

BOM. (2019). *Archive—forty-eight-monthly rainfall deficiency for Australia* (period—48 months, until 31 October 2009). Bureau of Meteorology. Published on 28 August 2019. http://www.bom.gov.au/jsp/awap/rain/archive.jsp?colour=colour&map=drought&year=2009&month=10&period=48month&area=nat. Accessed 8 January 2021.

BOM. (2020a, January 9). *Annual climate statement 2019*. Bureau of Meteorology. http://www.bom.gov.au/climate/current/annual/aus/2019/. Accessed 8 January 2021.

BOM. (2020b, December 10). *Archive - Six-monthly rainfall deficiency for Australia (period—6 months, until 31 December 2019)*. Bureau of Meteorology. http://www.bom.gov.au/jsp/awap/rain/archive.jsp?colour=col

our&map=drought&year=2019&month=12&period=6month&area=nat. Accessed 8 January 2020.

BOM. (2021, January 8). *Annual climate statement 2020*. Bureau of Meteorology. http://www.bom.gov.au/climate/current/annual/aus/. Accessed 8 January 2021.

Brears, R. (2017, January 17). Singapore transitioning towards urban water security. In R. Brears (Ed.), *Urban water security* (1st ed., pp. 225–241). John Wiley & Sons. https://doi.org/10.1002/9781119131755.ch13. Accessed 10 March 2020.

BSI. (n.d.). *Case study Gladstone area water board: 'Gladstone area water board (GAWB) are using ISO 55,001 certification to demonstrate the sound asset practices that contribute to the setting of water prices*. British Standards Institution. https://www.bsigroup.com/LocalFiles/EN-AU/_Case%20Studies/GAWB%20Case%20study-v5.pdf. Accessed 12 February 2020.

BSI. (2015). *Publicly available specification (PAS) 1192–5:2015 Specification for security-minded building information modelling, digital built environments and smart asset management*. British Standards Institution.

C3S. (2020, January 8). *Copernicus: 2019 was the second warmest year and the last five years were the warmest on record*. Press Release. European Copernicus Climate Change Service. https://climate.copernicus.eu/copernicus-2019-was-second-warmest-year-and-last-five-years-were-warmest-record. Accessed 8 January 2021.

C3S. (2021). *Copernicus: 2020 warmest year on record for Europe; globally, 2020 ties with 2016 for warmest year recorded*. https://climate.copernicus.eu/copernicus-2020-warmest-year-record-europe-globally-2020-ties-2016-warmest-year-recorded. Accessed 8 January 2021.

Canning, A., Deere, D., & Hill, K. (2020, January). *Bushfires and the risks to drinking water quality*. Water Research Australia. https://www.waterra.com.au/publications/fact-sheets/. Accessed 23 January 2020.

Close, H. (2020, February 27). Housebuilding ban on floodplains isn't enough—Flood-Prone communities should take back control. *The Conversation*. theconversation.com/housebuilding-ban-on-floodplains-isnt-enough-flood-prone-communities-should-take-back-control-132468. Accessed 3 March 2020.

Cole, G. (2011, October). *Time of use tariffs: Reforming the economics of urban water supply, waterlines report*. National Water Commission.

Cole, G., O'Halloran, K., & Stewart, R. A. (2012, January 12). Time of use tariffs: Implications for water efficiency. *Water Supply, 12*(1), 90–100.

https://iwaponline.com/ws/article-pdf/12/1/90/678710/90.pdf. Accessed 24 July 2020.

Defeo, J. A. (2017). *Juran's quality handbook* (7th ed.). McGraw-Hill Education.

Deloitte. (2019, June). *The social and economic cost of the North and Far North Queensland Monsoon Trough*. Prepared for Queensland Reconstruction Authority, Deloitte. https://www2.deloitte.com/content/dam/Deloitte/au/Documents/Economics/deloitte-au-dae-monsoon-trough-social-economic-cost-report-160719.pdf. Accessed 23 January 2020.

Doberstein, B., Fitzgibbons, J., & Mitchell C. (2019, August). Protect, accommodate, retreat or avoid (PARA): Canadian community options for flood disaster risk reduction and flood resilience. In *Natural Hazards: Journal of the International Society for the Prevention and Mitigation of Natural Hazards*, *98*(1), 31–50.

E + H. (n.d.). *Predykcyjny system sterowania Liquiline Control CDC81*. Endress + Hauser. https://www.pl.endress.com/pl/wieksza-wydajnosc-mniejsze-koszty/rozwiazania-analityczne/usuwanie-azotu/kontrola-napowietrzania-usuwanie-fosforu-w-oczyszczalni-sciekow. Accessed 20 July 2020.

E + H. (2018a). *Controlled by Liquiline Control Automated nitrogen removal in Stadtlohn wastewater treatment plant*. Endress + Hauser case study sheet. Endress + Hauser. https://www.pl.endress.com/pl/Endress-Hauser-Twoj-Partner/nasze-realizacje/stadthol-usuwanie-azotu. Accessed 21 July 2020.

E + H. (2018b). *Controlled by Liquiline Control. Automated phosphate removal in Stadtlohn wastewater treatment plant*. Endress + Hauser case study sheet. Endress + Hauser. https://www.pl.endress.com/pl/Endress-Hauser-Twoj-Partner/nasze-realizacje/automatyzacja_stracania_fosforu. Accessed 21 July 2020.

E + H. (2020). Jak zwiększyć efektywność i oszczędzać na kosztach oczyszczalni ścieków? Wywiad ze specjalista ZGWK w Tomaszowie Mazowieckim. *Kurier Wod-Kan. Magazyn klientów Endress + Hauser* (pp. 24–28).

Freeman, R. (2020, July 24). When to NOT use AI or use it based on my experience. *Towards Data Science online*. https://towardsdatascience.com/when-to-not-use-ai-or-use-it-based-on-my-experience-abb58c063aba. Accessed 8 July 2020.

Gelman, J. (2020, October 1). *Personal communication*. Kando.

GWP. (2019). *Water 4.0. Made in Germany*. German Water Partnership.

Glickman, J., & Leroi, A. (2015). *Adapt and adopt: Digital transformation for utilities*. Bain & Company. https://www.bain.com/insights/adapt-and-adopt-digital-transformation-for-utilities/. Accessed 8 June 2020.

Gourbesville, P. (2019). Smart water solutions for water security: From concept to operational implementation. In K. Lim, A. K. Makarigakis, O. Sohn, & B. Lee (Eds.) [(Editor-in-Chief)], *Water security and the sustainable development goals* (pp. 47–67). Global water security issues series. United Nations Educational, Scientific and Cultural Organization (UNESCO) International Centre for Water Security and Sustainable Management. https://unesdoc.unesco.org/ark:/48223/pf0000367904.locale=en. Accessed 18 March 2020.

Greenfield, D. (2020, June 4). A step approach to industry 4.0. *Automation World*. https://www.automationworld.com/factory/iiot/article/21135339/a-step-approach-to-industry-40. Accessed 18 June 2020.

Grievson, O. (2018, September 31). Water 4.0 and the wastewater cycle. *WWT Magazine*. Faversham House Group Ltd, 2020. https://wwtonline.co.uk/features/water-4-0-and-the-wastewater-cycle#.XJCgo7ixWM. Accessed 18 June 2020.

Grievson, O. (2020). *Is water 4.0 the future?* WEX global 2020. https://wex-global2019.com/is-water-4-0-the-future/. Accessed 18 June 2020.

Gulati, R. (2021). *Maintenance and reliability best practices* (3rd ed.). Industrial Press.

Hadjimichael, A., Comas, J., & Corominas, L. l. (2016, December 1). Do artificial intelligence methods enhance the potential of decision support systems? A review for the urban water sector. To appear in: AI Communications.

Hardwick, J., Killeen, M., Kohler, P., Lafraia, J., & Nugent, S. (2020, April 2). *Living Asset Management*. Living Asset Management Think Tank Incorporated.

Harken, B. A., & Brewster, M. M. (2019). Security and green growth: Supporting Development While Safeguarding Water Resources. In K. Lim, A. K. Makarigakis, O. Sohn, & B. Lee (Editor-in-Chief), *Water security and the sustainable development goals* (pp. 113–131). Global water security issues series. United Nations Educational, Scientific and Cultural Organization (UNESCO) International Centre for Water Security and Sustainable Management. https://unesdoc.unesco.org/ark:/48223/pf0000367904.locale=en. Accessed 18 March 2020.

Harris, C. (2018, December 5). Addressing non-revenue water losses. *Water Source*. https://watersource.awa.asn.au/business/assets-and-operations/addressing-non-revenue-water-losses/. Accessed 6 February 2020.

Hesamoddin, R., Khalilpour, K. R., Betts, J. M., & Tapper, N. (2019). Chapter 13—Energy-water nexus: Renewable-integrated hybridized desalination systems. In K. R. Khalilpour (Ed.), *Polygeneration with polystorage for chemical and energy hubs for energy and chemicals* (pp. 409–458). Academic Press (Elsevier Inc.). Published on 30 November 2018. https://www.sciencedirect.com/science/article/pii/B9780128133064000136. Accessed 6 February 2020.

Hewitson, B., Janetos, A. C., Carter, T. R., Giorgi, F., Jones, R. G., Kwon, W.-T., Mearns, L. O., Schipper, E. L. F., & van Aalst, M. (2014). Regional context. In V. R. Barros, C. B. Field, D. J. Dbokken, M. D. Mastrandrea, K. J. Mach, T. E. Bilir, M. Chatterjee, K. L. Ebi, Y. O. Estrada, R. C. Genova, B. Girma, E. S. Kissel, A. N. Levy, S. MacCracken, P. R. Mastrandrea, & L. L. White (Eds.), *Climate change 2014: Impacts, adaptation, and vulnerability—Part B: Regional Aspects* (pp. 1133–1197). Cambridge University Press. Contribution of Working Group II to the Fifth Assessment Report of the Intergovernmental Panel on Climate Change.

Honert, R. C. V., & McAneney, J. (2011). The 2011 Brisbane floods: Causes, impacts and implications. *Water, 3*, 1149–1173.

IET and Atkins. *Digital twins for the built environment.* Institution of Engineering and Technology. https://www.theiet.org/media/4719/digital-twins-for-the-built-environment.pdf. Accessed 3 September 2020.

IPART. (2020). *Sydney desalination plant expansion. Terms of reference.* IPART—the Independent Pricing & Regulatory Tribunal. Published on 8 January 2020. https://www.ipart.nsw.gov.au/Home/Industries/Water/Reviews/Metro-Pricing/Sydney-Desalination-Plant-Expansion?qDh=2. Accessed 11 March 2020.

IPCC. (2014). *Climate change 2014: Impacts, adaptation, and vulnerability. Part A: Global and sectoral aspects. Contribution of Working Group II to the Fifth Assessment Report of the Intergovernmental Panel on Climate Change* (C. B. Field, V. R. Barros, D. J. Dokken, K. J. Mach, M. D. Mastrandrea, T. E. Bilir, M. Chatterjee, K. L. Ebi, Y. O. Estrada, R. C. Genova, B. Girma, E. S. Kissel, A. N. Levy, S. MacCracken, P. R. Mastrandrea, & L. L. White, Eds., pp. 1–1131). Cambridge University Press

ISO. (2006). *ISO 14040:2006. Environmental management: Life cycle assessment—Principles and framework.* Geneva, Switzerland: International Organization for Standardization.

ISO. (2014a). *ISO 55000:2014 Asset management—Overview, principles and terminology.* International Organization for Standardization.

ISO. (2014b). *ISO 55001:2014 Asset management: Management systems—Requirements.* International Organization for Standardization.

ISO. (2015a). *ISO 9000:2015 Quality management systems—Fundamentals and vocabulary.* International Organization for Standardization.

ISO. (2015b). *ISO 9001:2015 Quality management systems—Quality management systems—Requirements*. International Organization for Standardization.

ISO. (2015c). *ISO 14001:2015 Environmental management systems—Requirements with guidance for use*. International Organization for Standardization.

ISO. (2018a). *ISO 22000:2018 Food safety management systems—Requirements for any organization in the food chain*. International Organization for Standardization.

ISO. (2018b). *ISO 31000:2018 Risk management—Guidelines*. Geneva, Switzerland: International Organization for Standardization.

ISO. (2018c). *ISO 55002:2018 Asset management—Management systems—Guidelines for the application of ISO 55,001*. International Organization for Standardization.

ISO. (2018d). *ISO 45001:2018 Occupational health and safety management systems—Requirements with guidance for use*. International Organization for Standardization.

ISO. (2018e). *Asset Management: Achieving the UN Sustainable Development Goals*. International Organization for Standardization, Technical Committee 251 (ISO/TC 251).

ISO. (2018f). *ISO 19650–1:2018 Organization and digitization of information about buildings and civil engineering works, including building information modelling (BIM): Information management using building information modelling—Part 1: Concepts and principles*. International Organization for Standardization.

ISO. (2018g). *ISO 19650–2:2018 Organization and digitization of information about buildings and civil engineering works, including building information modelling (BIM)—Information management using building information modelling—Part 2: Delivery phase of the assets*. International Organization for Standardization.

ISO. (2018h). *ISO/IEC 27000:2018 Information technology—Security techniques—Information security management systems—Overview and vocabulary*. International Organization for Standardization.

ISO. (2019a). *ISO/TS 55010:2019 Asset management—Guidance on the alignment of financial and non-financial functions and asset management*. International Organization for Standardization.

ISO. (2019b). *ISO/TC 251 Asset management: ISO/TS 55010:2019 asset management—Guidance on alignment of asset management, finance and accounting*. International Organization for Standardization. https://committee.iso.org/sites/tc251/home/projects/published/isots-55010.html. Accessed 29 November 2020.

ISO. (2020a). *ISO 19650–3:2020: Organization and digitization of information about buildings and civil engineering works, including building information modelling (BIM)—Information management using building information modelling—Part 3: Operational phase of the assets*. International Organization for Standardization.

ISO. (2020b). *ISO 19650–5:2020: Organization and digitization of information about buildings and civil engineering works, including building information modelling (BIM)—Information management using building information modelling—Part 5: Security-minded approach to information management*. International Organization for Standardization.

Jacob, D. (2017). *Quality 4.0 Impact and strategy handbook: Getting digitally connected to transform quality management*. LNS Research. https://www.sas.com/content/dam/SAS/en_us/doc/whitepaper2/quality-4-0-impact-strategy-109087.pdf. Accessed 8 June 2020.

JBA Group. (2019, March 7). *One month on: A retrospective look at the Townsville flooding in February 2019*. https://www.jbarisk.com/flood-services/event-response/a-retrospective-of-townsville-flooding/. Accessed 23 January 2020.

Kando. (2020a, June 15). *Kando and Paperspace Partner to Bring Advanced Machine Learning to Municipal Systems Monitoring*. https://blog.paperspace.com/kando-and-paperspace-partner-to-bring-advanced-machine-learning-to-municipal-systems-monitoring/. Accessed 22 July 2020.

Kando. (2020b). *Sensors and smart analytics for tracking illicit sewer connections hotspots*. https://www.digital-water.city/solution/sensors-and-smart-analytics-for-tracking-illicit-sewer-connections-hotspots/. Accessed 22 July 2020.

Kando (2020c, June). *Kando customer success stories*.

Kennedy, J. (2017). Asset management—Defending the business. *The Asset Journal*, *11*(3), 4–12. Asset Management Council.

Kennedy, R. F. (1964, May 25). *Address by Attorney General Robert F. Kennedy before the United States Conference of Mayors*. U.S. Department of Justice. https://www.justice.gov/ag/rfkspeeches/1964/05-25-1964.pdf. Accessed 26 December 2020.

Kolditz, O., Karsten Rink, K., Nixdorf, E., Fischer, T., Bilke, L., Naumov, D., Liao, Z., & Yue, T. (2019). Environmental information systems: Paving the path for digitally facilitated water management (Water 4.0). *Engineering, 5*, 828–832.

Kovats, R. S., Valentini, R., Bouwer, L. M., Georgopoulou, E., Jacob, D., Martin, E., Rounsevell, M., & Soussana J.-F. (2014). Europe. In *Climate change 2014: Impacts, adaptation, and vulnerability. Part B: Regional Aspects.*

(V. R. Barros, C. B. Field, D. J. Dbokken, M. D. Mastrandrea, K. J. Mach, T. E. Bilir, M. Chatterjee, K. L. Ebi, Y. O. Estrada, R. C. Genova, B. Girma, E. S. Kissel, A. N. Levy, S. MacCracken, P. R. Mastrandrea, & L. L. White, Eds., pp. 1267–1326). Cambridge University Press. Contribution of Working Group II to the Fifth Assessment Report of the Intergovernmental Panel on Climate Change.

Lall, U. (2019). Positive water sector disruptions by 2030, III. In F. Machado & L. M. Mimmi (Eds.), *The future of water a collection of essays on "disruptive" technologies that may transform the water sector in the next 10 years* (pp. 45–61). Inter-American Development Bank, Water and Sanitation Division.

Lipnicki, P. (2021, January 5). *Personal communication.* ReliaSol.

Lloret, J., TomásGironés, J., Canovas Solbes, A., & Parra-Boronat, L. (2016). An integrated IoT architecture for smart metering. *IEEE Communications Magazine, 54*(12), 50–57.

Lovelock, J. (2019, July 4). *Novacene: The coming age of hyperintelligence* (1st ed.). Penguin Books Limited.

Lubbers, C. L., Icke, O., van Eijden, R., de Koning, M., Huising C., & de Wit R. (2018, July 17–18). *Performance improvement of wastewater transport systems and treatment processes by performance monitoring and predictive control.* In the Proceedings of 12th European Waste Water Management Conference, Aqua Environment, MA, UK.

Maltese Island Weather. (2020, January 5). *End of year report. 2019 in review.* https://www.malteseislandsweather.com/2019-the-warmest-year-this-decade/. Accessed 5 January 2021.

Maltese Island Weather. (2021, January 4). *End of year report. 2020 in review.* https://www.malteseislandsweather.com/2020-an-overall-warm-and-dry-year/. Accessed 5 January 2021.

McKinsey. (2020). *An executive's guide to AI*. McKinsey & Company. https://www.mckinsey.com/business-functions/mckinsey-analytics/our-insights/an-executives-guide-to-ai. Accessed 8 December 2020.

Mechefske, C. K. (2005). Machine condition monitoring and fault diagnosis, Chapter 25. In C. W. de Silva (Ed.), *Vibration and Shock handbook.* CRC Press, Taylor and Francis Group.

Mitchell, J. S. (1978, December 10–15). *Bearing diagnostics: An overview* (pp. 15–24). Winter Annual Meeting. American Society of Mechanical Engineers.

Microsoft. (2016, October). *Water industry: Benefits of moving to Cloud technology.* Prepared for Water Sector Organisations, United Kingdom.

https://info.microsoft.com/rs/157-GQE-382/images/WaterSectorCloudPoV.PDF. Accessed 8 July 2020.

National Statistics Office. (2019). *Key figures for Malta. 2019 Edition. Visuals & words.* National Statistics Office. https://nso.gov.mt/en/nso/Media/Salient-Points-of-Publications/Documents/Key%20Figures%20for%20Malta%20-%202019%20Edition/Malta%20In%20Figures%20-%202019.pdf. Accessed 18 March 2020.

Newhart, K. B., Holloway, R. W., Hering, A. S., & Cath, T. Y. A. (2019, March 21). Data-driven performance analyses of wastewater treatment plants: A review. *Water Research,* 157, 498–513.

Nguyen, H., Mehrotra, R., & Sharma, A. (2020, January). Assessment of climate change impacts on reservoir storage reliability, resilience and vulnerability using a Multivariate Frequency Bias Correction approach. *Water Resources Research.* Accepted on 15 January 2020. https://agupubs.onlinelibrary.wiley.com/doi/abs/10.1029/2019WR026022. Accessed 12 February 2020.

NHMRC. (2018). *National Water Quality Management Strategy. Australian Drinking Water Guidelines 6 2011.* Version 3.5 updated on August 2018. Canberra, Australia: National Health and Medical Research Council.

NRMMC, EPHC, and AHMC. (2006). *National Water Quality Management Strategy. Australian Guidelines 21 for Water Recycling. Managing Health and Environmental Risks.* Canberra, Australia: Natural Resource Management Ministerial Council, Environment Protection and Heritage Council, Australian Health Ministers Conference.

NOAA. (2020a). *What are El Niño and La Niña?* National Ocean Service. Last updated on 10 February 2020. https://oceanservice.noaa.gov/facts/ninonina.html. Accessed 12 February 2020.

NOAA. (2020b, January 3). *Catastrophic wildfires in southeastern Australia in 2019–20.* Washington, DC, United States: National Ocean Service. www.climate.gov/news-features/event-tracker/catastrophic-wildfires-southeastern-australia-2019-20. Accessed 12 February 2020.

Ofwat. (2007, November 13). *Providing best practice guidance on the inclusion of externalities in the ELL calculation, main report v05.* Ref: PROC/01/0075, Birmingham, United Kingdom: Office of Water Services. Licensed under the "Open Government Licence".

Ofwat. (2020, May). *Reference of the PR19 final determinations: Cost efficiency—response to common issues in companies' statements of case* (p. 56). Birmingham, United Kingdom: Office of Water Services. Licensed under the "Open Government Licence".

Passos, C. (2020, July 16). *Personal communication.* Dynamox.

Pfeiffer, S. (2017). The vision of "industrie 4.0" in the making—A case of future told, tamed, and traded. *Nanoethics, 11*, 107–121. Springer Nature B.V. Published online on 25 January 2017. https://link.springer.com/article/10.1007/s11569-016-0280-3. Accessed 19 August 2020.

Piotrowski, J. (2007). *Shaft alignment handbook* (3rd ed.). CRC Press, a Taylor & Francis Group.

Poljak, D. (2018). *Industry 4.0—New challenges for public water supply organizations.* https://bib.irb.hr/datoteka/940425.INDUSTRY_4.0__New_Challenges_for_Public_Water_Supply_Organizations.pdf. Accessed 8 July 2020.

Rabiee, H., Khalilpour, K. R., Betts, J. M., & Tapper, N. (2019). Chapter 13, Energy-water nexus: Renewable-integrated hybridized desalination systems. In K. R. Khalilpour (Ed.), *Polygeneration with polystorage for chemical and energy hubs for energy and chemicals* (pp. 409–458). London: Academic. Available online on 30 November 2018. www.sciencedirect.com/science/article/pii/B9780128133064000136. Accessed 6 February 2020.

Rayner, R. (2019, January). Chapter 7—Incorporating climate change within Asset Management. In C. Lloyd & M. Corcoran. *Asset management* (2nd ed., pp. 143–161). ICE Publishing, Institution of Civil Engineers.

Reddy, R. (2020, June 17). *Machine learning 101 in predictive maintenance. IndustryWeek.* https://www.industryweek.com/technology-and-iiot/article/21134278/machine-learning-101-in-predictive-maintenance. Accessed 8 July 2020.

Reisinger, A., Kitching, R. L., Chiew, F., Hughes, L., Newton, P. C. D., Schuster, S. S., Tait, A., & Whetton P. (2014). Australasia. In *Climate change 2014: Impacts, adaptation, and vulnerability. Part B: Regional Aspects* (V. R. Barros, C. B. Field, D. J. Dokken, M. D. Mastrandrea, K. J. Mach, T. E. Bilir, M. Chatterjee, K. L. Ebi, Y. O. Estrada, R. C. Genova, B. Girma, E. S. Kissel, A. N. Levy, S. MacCracken, P. R. Mastrandrea, & L. L. White, Eds., pp. 1371–1438). Cambridge University Press. Contribution of Working Group II to the Fifth Assessment Report of the Intergovernmental Panel on Climate Change.

RHDHV. (2020a). Aquasuite. White paper: *Introducing* the water industry's virtual *operator which never sleeps. Freeing up your operational team's time for high value tasks.* Royal HaskoningDHV. https://analytics-eu.clickdimensions.com/royalhaskoningdhvcom-agyu4/pages/m0sryp6feeqiwbqvrccvq.html?utm_source=Water-for-Industry-securing-business-continuity&utm_medium=AI-POWERED-VIRTUAL-OPERATOR&utm_campaign=Water-for-Industry. Accessed 23 July 2020.

RHDHV. (2020b, January/February). *Aquasuite PURE.* Royal HaskoningDHV. https://aquasuite.ai/en/products/pure/. Accessed 23 July 2020.

RHDHV. (2020c). *Achieving automated operations and data driven insights from water reuse*. waterwastewaterasia.com. Royal HaskoningDHV (pp. 24–25). https://waterwastewaterasia.com/january-february-2020/. Accessed 23 July 2020.

Risk Society. *Disaster risk assessment through the prism of ISO:31000*. The Hague. Risk Society. https://www.risk-society.com/bibliotheek/blog1/151-disaster-risk-assessment-throughthe-prism-of-iso-31000. Accessed 13 February 2020.

Rodriguez & Sons Pty Ltd v. (2019). Queensland Bulk Water Supply Authority and Lynette Joy Lynch v. Queensland Bulk Water Authority trading as Seqwater. 2014/200854 and 2016/373183, New South Wales Supreme Court, Court judgement delivered on 29 November 2019. www.supremecourt.justice.nsw.gov.au/Pages/sco2_classaction/floods.aspx. Accessed 17 January 2020.

Roorda, J. (2020, January 27). Bushfires—Fighting back with an Asset Management Plan. *Talking Infrastructure*. http://talkinginfrastructure.com/. Accessed 28 January 2020.

Sapiano, M. (2018, August 30). *Interview—Malta: Water scarcity is a fact of life*. Copenhagen, Denmark: European Environment Agency. Published 30 2018. https://www.eea.europa.eu/signals/signals-2018-content-list/articles/interview-2014-malta-water-scarcity. Accessed 3 February 2020.

Sarni, W. (2019). The future of water is digital, IV. In F. Machado & L. M. Mimmi (Eds.), *The future of water a collection of essays on "disruptive" technologies that may transform the water sector in the next 10 years* (pp. 62–75). Inter-American Development Bank, Water and Sanitation Division.

Sedlak, D. (2015). *Water 4.0: The past, present, and future of the world's most vital resource*. Yale University Press. Reprint Edition, 31 March 2015.

Seqwater. (n.d.). *Fact sheet: Realities of rain—What is a flood?* Seqwater. https://yourseqwater.com.au/37596/documents/86232. Accessed 3 February 2020.

Seqwater. (2017a, March). *Queensland bulk water supply authority, trading as Seqwater: Water for life South East Queensland's Water Security Program 2016–2046*. Version 2. Seqwater.

Seqwater. (2017b, July 31). *2018 bulk water price review. Seqwater Submission. Part A*. Seqwater.

Seqwater. (2018). *Seqwater Water Grid map as at 30 June 2018*. Seqwater. https://www.seqwater.com.au/sites/default/files/2019-08/48125%20-%

20SEQ%20Water%20Grid%20map%20-%20AS%20AT%2030%20JUNE%202018_0.pdf. Accessed 3 February 2020.

Seqwater. (2019, May 21). *Lessons from the millennium drought*. Seqwater. https://www.seqwater.com.au/news/lessons-millennium-drought. Accessed February 2020.

SKF. (2020). *AI for wastewater treatment Big Data's untapped potential for Water/Wastewater*. https://industrial-ai.skf.com/ai-for-wastewater-treatment/. Accessed 8 June 2020.

Smart Water Magazine. (2020a). The digital twin, another step forward in ACCIONA's commitment to technology. *Smart Water Magazine*, SWM Monthly 3. https://smartwatermagazine.com/blogs/julio-de-la-rosa/digital-twin-another-step-forward-accionas-commitment-technology. Accessed 7 August 2021.

Smart Water Magazine. (2020b). Technology against climate change: ACCIONA's solutions for more efficient management. *Smart Water Magazine*, SWM Monthly 1. https://smartwatermagazine.com/news/acciona/technology-against-climate-change-accionas-solutions-more-efficient-management. Accessed 10 July 2020.

Stankovic, M., Hasanbeigi, A., Neftenov, N., & Tambourine Innovation Ventures. (2020, April). *Use of 4IR technologies in water and sanitation in Latin America and the Caribbean*. Inter-American Development Bank, Water and Sanitation Division.

Steutel-Maron, A. (2019a, December 2). *How Big Data enables smart collection systems and protects Wastewater Treatment Plants*. https://medium.com/datadriveninvestor/how-big-data-enables-smart-collection-systems-and-protect-wastewater-treatment-plants-32595fbe6852. Accessed 22 July 2020.

Steutel-Maron, A. (2019b, January 17). *How to digitize your wastewater network with a smart water solution?*. https://medium.com/@annelisteutel/how-to-digitize-your-wastewater-network-with-a-smart-water-solution-1eebffb1b836. Accessed 22 July 2020.

Steward, K. (2020, July 14). Real time Wastewater Monitoring Enables Rapid COVID-19 Outbreak Detection. *Industry Insight*. https://www.technologynetworks.com/applied-sciences/blog/realtime-wastewater-monitoring-enables-rapid-covid-19-outbreak-detection-337391. Accessed 22 July 2020.

Świerczewska, M. (2020, August 7). *Personal communication*. Endress + Hauser.

Taggle Systems. (2019). https://taggle.com/. Accessed 19 August 2020.

UN. (1997). *Glossary of environment statistics.* Studies in Methods, Series F, No. 67, New York, N.Y., Unites States: Department for Economic and Social Information and Policy Analysis, Statistics Division, United Nations.

UN. (n.d.). https://www.un.org/sustainabledevelopment/sustainable-development-goals/. United Nations. Accessed 19 August 2020.

Vestner, R., & Keilholz, P. (2016). Was bedeutet der „4.0-Ansatz " für die Wasserwirtschaft? *Gewässerschutz Wasser Abwasser, Bd, 239*, 7/1 – 7/13.

Voutchkov, N. (2019). Disruptive innovations in the water sector, II. In F. Machado & L. M. Mimmi (Eds.), *The future of water a collection of essays on "disruptive" technologies that may transform the water sector in the next 10 years* (pp. 14–44). Inter-American Development Bank, Water and Sanitation Division.

Water Corporation. (2013, August). *Groundwater replenishment scheme communications strategy 2013–2016*. https://consultation.epa.wa.gov.au/seven-day-comment-on-referrals/perth-groundwater-replenishment-scheme/supporting_documents/Appendix%2012%20Communication%20Strategy%202013%202016.pdf. Accessed 7 August 2021.

Water Services Corporation. (n.d.). *New water.* http://www.wsc.com.mt/information/new-water/. Accessed 15 January 2020.

Water Services Corporation. (2019, March 8). *Fresh water supply from Malta to Gozo restored*. http://www.wsc.com.mt/fresh-water-supply-from-malta-to-gozo-restored/. Accessed 15 January 2020.

Water Source. (2018). Seqwater pouring $1 million into protecting Brisbane River. *Water Source*. Australian Water Association, Published on 12 October 2018. https://watersource.awa.asn.au/environment/natural-environment/seqwater-pouring-1-million-into-protecting-brisbane-river/. Accessed 30 January 2020.

Water Source. (2019). How did the 2019 Townsville floods affect the city's water infrastructure? An on the ground look at the 1 in 500 year monsoon rain event. *Water Source*. Australian Water Association, Published on 24 April 2019. https://watersource.awa.asn.au/business/assets-and-operations/how-did-the-2019-townsville-floods-affect-the-citys-water-infrastructure/. Accessed 29 January in 2020.

Water Source. (2020, January 8). Mobile desal plants to treat brackish water in regional NSW. *Water Source*. Australian Water Association. https://watersource.awa.asn.au/community/engagement/mobile-desal-plants-to-treat-brackish-water-in-regional-nsw/. Accessed 29 January in 2020.

Watkinson, A., Volders, A., Smolders, K., Simms, A., Olley, J., Burford, M., Stratton, H., Gibbes, B., & Grinham, A. (2012, April). Source water protection for SEQWater. Novel techniques to assess the effectiveness of management intervention and prioritise action. Sydney, Australia: Australian Water Association. *Water, 30*(2), 100–105.

WEF. (2017). *Intelligent water systems: The path to a smart utility*. Water Environment Federation (WEF). WSEC-2016-WP-002.

WEF. (2018, September). *Harnessing the fourth industrial revolution for water*. Fourth Industrial Revolution for the Earth Series. Geneva, Switzerland: World Economic Forum. https://www.weforum.org/reports/harnessing-the-fourth-industrial-revolution-for-water. Accessed 19 June 2020.

Weitze, L., Solas, H., & Palzer, U. (2018, October 18). *Digital assessment to increase efficiency in the water infrastructure—Water 4.0*. Weimar Institute of Applied Construction Research. A Presentation from German American Dialogue for Sustainable Water Infrastructure and Technologies in Milwaukee/Wisconsin, United States. https://www.researchgate.net/publication/328502279_Digital_Assessment_to_Increase_Efficiency_in_the_Water_Infrastructure_-_Water_40. Accessed 9 July 2020.

Wikipedia. https://en.wikipedia.org/wiki/List_of_desalination_plants_in_Australia. Accessed 27 February in 2020.

Williams, W. (2017, November 10). *U.S. utilities weigh the cost and benefit of ISO 55,001 certification: Breaking energy*. https://breakingenergy.com/. Accessed 29 January in 2020.

Wong-Parodi, G., Fischhoff, B., & Strauss, B. (2015). Resilience vs. adaptation: Framing and action. *Climate Risk Management, 10*, 1–7. Elsevier B.V.

World Economic Forum. (2020). *The global risks report 2020*. Insight Report. 15th Edition. Geneva, Switzerland: World Economic Forum in partnership with Marsh & McLennan and Zurich Insurance Group. http://www3.weforum.org/docs/WEF_Global_Risk_Report_2020.pdf. Accessed 18 March 2020.

Youssef, Y. W., & Khodzinskaya, A. (2019, May 29). *A review of evaporation reduction methods from water surfaces*. In E3S Web of Conferences 97, 05,044 (2019), XXII International Scientific Conference *Construction the Formation of Living Environment*. https://www.e3s-conferences.org/articles/e3sconf/abs/2019/23/e3sconf_form2018_05044/e3sconf_form2018_05044.html. Accessed 29 January in 2020.

Xiupeng, W. (2013). *Modeling and optimization of wastewater treatment process with a data-driven approach* (PhD Doctor of Philosophy, thesis). University of Iowa, USA. https://iro.uiowa.edu/discovery/fulldisplay/alma9983777011302771/01IOWA_INST:ResearchRepository. Accessed 12 July 2020.

Xu, Z., Zhang, L., Zhao, L., Li, B., Bhatia, B., Wang, C., Wilke, K. L., Song, Y., Labban, O., Lienhard, J. H., Wang, R., & Wang, E. N. (2020). Ultrahigh-efficiency desalination via a thermally-localized multistage solar still. In *Energy & environmental science*. Royal Society of Chemistry. Accepted on 15 January 2020. https://doi.org/10.1039/c9ee04122b. https://pubs.rsc.org/en/content/articlelanding/2020/EE/C9EE04122B#!divAbstract. Accessed 13 February 2020.

Xylem. (n.d.). *Harness the power of decision intelligence: Driving performance improvement with smarter water, a brochure*. https://www.xylem.com/en-us/campaigns/act/decision-intelligence/. Accessed 10 November 2020.

Xylem. (2018). *Clean water solutions and services: Total control throughout the clean water cycle, a brochure*. https://www.xylem.com/siteassets/industries--applications/resources/xylem-en-uk-clean-water-brochure_en.pdf. Accessed 10 November 2020.

Xylem. (2019). *Cuxhaven, Germany. Reducing aeration energy usage by 26%, minimizing operational expenses, and reducing compliance risk by applying a decision intelligence approach*, a case study leaflet. https://www.xylem.com/en-us/support/case-studies-white-papers/cuxhaven-germany-reducing-aeration-energy-usage-minimizing-operational-expenses-and-reducing-compliance-risk-by-applying-a-decision-intelligence-approach/. Accessed 10 November 2020.

Xylem. (2020a). *Real-time decision support helps cut Combined Sewer Overflow volume by 100 million gallons and reduces capital needed for regulatory compliance. Evansville Water and Sewer Utility, Evansville, IL, USA*. https://www.xylem.com/en-us/support/case-studies-white-papers/real-time-decision-support-helps-cut-combined-sewer-overflow-volume-by-100-million-gallons-and-reduces-capital-needed-for-regulatory-compliance/. Accessed 10 November 2020.

Xylem. (2020b). *Sensus, a Xylem brand*. https://sensus.com/communication-networks/sensus-technologies/flexnet-international/. Accessed 10 November 2020.

Xylem. (2020c). *White paper: Optimizing leak detection using virtual district metering areas (DMAs).* https://www.xylem.com/en-us/support/case-studies-white-papers/optimizing-leak-detection-with-a-virtual-dma/. Accessed 10 November 2020.

Ziemer, C., & Volker, C. (2017, September 1). Water 4.0: What it Means for the German Water Industry. *WaterWorld*. https://www.waterworld.com/international/utilities/article/16201159/water-40-what-it-means-for-the-german-water-industry. Accessed 8 July 2020.

Index

A

Adaptability 10, 143, 152, 153
Adaptation 29, 58
Advanced Meter Infrastructure (AMI) 120, 122
Advanced Water Treatment Plants 38
Aquasuite 95
Artificial Intelligence (AI) 10, 56
Asset 10
Asset Information Management (AIM) 149
Asset management 10, 143, 149
Asset Management 4.0 75, 147
Asset management fundamentals 142
Asset Management Plan (AMP) 11, 144
Asset Management Systems (AMS) 5, 11, 38, 68, 78
Australia 30, 31
  drought and megafires in 37
Australian Drinking Water Guidelines (ADWG) 62
Australian Guidelines for Water Recycling 62
Automated Meter Management (AMM) 64
Automated Meter Reading (AMR) Systems 120

B

Big data 75, 82, 147
BLU-X 97
Building Information Modelling (BIM) 149
Built assets 11

C

CapEx (Capital Expenditures) 152

R. Kijak, *Water Asset Management in Times of Climate Change and Digital Transformation*, Palgrave Studies in Climate Resilient Societies,
https://doi.org/10.1007/978-3-030-79360-9

Carbon footprints 5, 104
Clear Upstream 111
Climate adaptation 12
Climate change 28
  risk-based approach to 28
Climate resilience 5, 12, 30
  and adaptation 30
Climate risks 54
Combined Sewer Overflows (CSO) 12, 56, 81
Computerised Maintenance Management System (CMMS) 79
Continuous improvement 12
Cyberattacks 83
Cyber-Physical System (CPS) 13, 80, 149
Cyber-Physical Water Systems (CPWS) 80
Cybersecurity 83, 154

D

Deming's Plan-Do-Check-Act (PDCA) 145, 147
Desalinated seawater 64
Desalinated water 62
Desalination plants 55
Digital transformation 2, 4
Digital twins 13, 57, 75, 80, 82, 88, 147, 149
Drought 14, 37, 55
Dynamox (Brazil) 107
DynaPredict solution 105

E

Economic Level of Leakage (ELL) 14, 59, 63
El Niño-Southern Oscillation (ENSO) 15, 35, 36
Endress+Hauser (Germany) 90
Engineering and business processes 4.0 75
Evidence-based decision making 147
Exposure 15, 29

F

Financial and non-financial functions, alignment of 151
Fires 55
Flood-drought cycle 37
Floods 55
Fourth industrial revolution 2
Fourth water revolution 3
  scope and timescales of 75

G

Gold Coast Desalination Plant 38

H

Hazard Analysis and Critical Control Points (HACCP) 16, 62
Hazards 16, 29
Horizontal integration 5, 78
Hybrid analytics 17, 88

I

Improvement 152
Industrial Internet of Things (IIoT) 18
Industry 4.0 2, 4, 74, 78
Information security 154
Information Security Management System (ISMS) 154

Infrastructure Leakage Index (ILI) 17, 59
Integrated business model 147
Intergovernmental Panel on Climate Change (IPCC)
Fifth Assessment Report 31, 56
Internet of Things (IoT) 18, 81
connected sensors 88
ISO 19650 149
ISO 19650-5 154
ISO 31000 risk management standard 29
ISO 55001 5, 149
ISO 55001 based Asset Management System 75, 142, 147, 148
ISO 55002 154
ISO 55010 152
ISO/IEC 27000 154

K

Kando (Israel) 111

L

Leadership culture 143
LeakView solutions 120
Level of Service (LOS) 2, 18
impact 44
Life Cycle Assessment (LCA) 18
Liquiline Control System 90

M

Machine Learning (ML) 19, 57
Maintenance 4.0 74
Malta 30, 31
Mean Time to Failure (MTTF) 79

N

Natural assets 19, 66
Neural Networks (NN) 10, 19, 100
New water 2, 22, 59, 63
Non-Revenue Water (NRW) 19, 55, 59, 63, 64

O

1999-2010 Millennium Drought 38
OpEx (Operational Expenditures) 152
Organisational culture 143
Overall Equipment Effectiveness (OEE) 4, 20, 56, 142, 156

P

PARA framework 55, 59
People–processes–technology 147
Predictive analytics 147
Programmable Logic Controller (PLC) 20
Project Information Management (PIM) 149
PUB (Singapore) 122
Purified Recycled Water (PRW) 2, 20, 22, 31, 38, 55, 59, 62, 64

Q

Quality 4.0 75
Queensland 30, 37, 38, 40

R

Recycled water 2, 62
ReliaSol (Poland) 100
Remaining Useful Life (RUL) 79

Resilience 29, 58
Reverse osmosis (RO) 60
Risk management 29
Risk treatment 29
Riverine flooding 20
Root Cause Failure Analysis (RCFA) 79
Royal HaskoningDHV (the Netherlands) 91

S

Seawater desalination 31, 58, 60
Security minded management 154
Smart meters 64
Smart water meters 88, 119
Smart water networks 88, 119–121
Source-to-tap 81
Southern Europe 30
Stormwater harvesting 31
Strategic Asset Management Plan (SAMP) 11, 143
Supervisory Control and Data Acquisition (SCADA) 20, 81, 88, 101
SurgeView solutions 120
Sustainable Development Goals (SDGs) 4, 28

T

Time of Use Tariffs (TOUT) 119
TotEx (Total Expenditures) 20, 152
2011 Queensland Flood 40
2019 Flood in Townsville (North Queensland) 43
2019-2020 *megafires* 67

V

vDMA 123
Vertical integration 5, 78
Virtual District Metering Areas (vDMAs) 121
Virtual operator 91
Virtual sensors 20
Vulnerability 20, 29

W

Waste 21
Wastewater Treatment Plant (WWTP)
  bypasses 56, 81
Water 1.0 76
Water 2.0 76
Water 3.0 76
Water 4.0 3, 21, 74, 75, 79, 80, 82, 147
  adaptation 5, 6
  opportunities and risks 81
  technologies and solutions 78, 88
Water footprints 5, 104
Water quality 45
Water quantity 44
Water recycling 58, 60
Water Security Program 38
Water storages 64
Water supply assets 22
  design and operational considerations for 53
Water transmission and distribution 60
Water treatment, transmission and distribution 59

Wide Bay Water (WBW) (Queensland, Australia) 121

X

Xylem's View solutions 120
Xylem (United States) 123

The manufacturer's authorised representative in the EU is Springer Nature Customer Service Centre GmbH, Europaplatz 3, 69115 Heidelberg, Germany. If you have any concerns regarding our products, please contact ProductSafety@springernature.com

Printed and bound by CPI Group (UK) Ltd, Croydon, CR0 4YY

30/09/2026

02225700-0001